Dietmar Korn

Silicon-Organic Hybrid Platform for Photonic Integrated Circuits

Karlsruhe Series in Photonics & Communications, Vol. 15
Edited by Profs. J. Leuthold, W. Freude and C. Koos

Karlsruhe Institute of Technology (KIT)
Institute of Photonics and Quantum Electronics (IPQ)
Germany

Silicon-Organic Hybrid Platform for Photonic Integrated Circuits

by
Dietmar Korn

KIT Scientific Publishing

Dissertation, Karlsruher Institut für Technologie (KIT)
Fakultät für Elektrotechnik und Informationstechnik, 2013

Impressum

KIT Scientific Publishing

Karlsruher Institut für Technologie (KIT)
KIT Scientific Publishing
Straße am Forum 2
D-76131 Karlsruhe

KIT Scientific Publishing is a registered trademark of Karlsruhe Institute of Technology. Reprint using the book cover is not allowed.

www.ksp.kit.edu

Print on Demand 2015

ISSN 1865-1100
ISBN 978-3-7315-0430-6
DOI 10.5445/KSP/1000048792

Silicon-Organic Hybrid Platform
for
Photonic Integrated Circuits

Zur Erlangung des akademischen Grades eines

DOKTOR-INGENIEURS

der Fakultät für Elektrotechnik und Informationstechnik
des Karlsruher Instituts für Technologie

genehmigte

DISSERTATION

von

Dipl.-Phys. Dietmar Korn

aus

Potsdam

Tag der mündlichen Prüfung:	02.10.2013
Hauptreferent:	Prof. Dr. sc. nat. Juerg Leuthold
Korreferent:	Prof. Dr.-Ing. Dr. h.c. Wolfgang Freude

Table of Contents

Zusammenfassung

Die vorliegende Arbeit erkundet das technische Potential einer neuen Plattform für hochintegrierte Optik. Diese Plattform kombiniert ausgewählte organische Materialien mit optischen, integrierten Schaltungen der Silizium-Photonik zu silizium-organischen Hybrid-Bauteilen (SOH) mit außergewöhnlichen Eigenschaften, d.h. Eigenschaften, welche Silizium allein nicht besitzt. Es werden Schlüsselkomponenten wie Laser und Modulatoren konzipiert, mit CMOS-basierten Technologien[1] hergestellt und charakterisiert. Dabei werden einzigartige Vorteile und praxisrelevante Eigenschaften der SOH-Bauteile identifiziert und schließlich mit Blick auf entsprechende Anwendungsszenarien demonstriert.

Die Miniaturisierung in der Elektronik durch die Entwicklung der auf Silizium beruhenden CMOS-Technologie hat entscheidend zur Bildung der heutigen Informationsgesellschaft beigetragen. Unsere Gesellschaft ist auf die Verarbeitung und auch den Transport von großen Datenmengen angewiesen. Nur durch die immer effizientere Nutzung von Ressourcen wie Energie und Rohstoffen lässt sich diese Entwicklung auch weiterführen [1]. Insbesondere die Infrastruktur des Internets, aber auch Analysemethoden und Sensorik, welche immer umfangreichere Daten liefern sollen, lassen sich mit Hilfe optischer Methoden weiter verbessern.

Für den Ausbau des Internets ist es einerseits wichtig, die Kapazität der optischen Langstreckenübertragungswege durch z.B. die Verwendung von komplexen Modulationsformaten und spektral effizienten Multiplexverfahren auszubauen. Andererseits ist es erforderlich, den Energieverbrauch der Datenzentren zu verringern. Dies betrifft nicht nur die Datenspeicher und Prozessoren, sondern insbesondere auch den Transport von Daten innerhalb der Datenzentren selbst [2]. Der Flaschenhals in der Informationsverarbeitung ist oftmals die Kommunikation zwischen den Servern. Die Umstellung auf optische Verbindungen ist bereits in vollem Gange. Allerdings verlangen Verbindungslängen von z.T. mehr als 300 m und der Bedarf nach höherer

[1] *Engl.* complementary metal oxide semiconductor, Halbleiterprozesse zur Fabrikation von gleichnamigen Halbleiterbauelementen in integrierten elektrischen Schaltkreisen für Computer und andere Geräte.

Bandbreite nach Alternativen zu bereits verwendeten optischen Systemen mit direkt modulierten, vertikal emittierenden Lasern (*engl.* vertical surface emitting laser, VCSEL). Die Entscheidung von Google oder Facebook, ihre Server-Farmen an kalten Orten mit kostengünstiger Energieversorgung zur Bewältigung des Kühlproblems zu bauen, unterstreicht die Feststellung, dass ein Wachstum der Internetinfrastruktur nur mit deutlicher Verbesserung der Energieeffizienz möglich ist. Wenn es gelänge, die notwendigen elektronischen integrierten Schaltkreise (engl. integrated circuit, IC) und die photonischen integrierten Schaltkreise (engl. photonic integrated circuit, PIC) mit ein und derselben Plattform in räumlicher Nachbarschaft, vielleicht sogar auf demselben Substrat unterzubringen, dann ließen sich Kosten und Energieverbrauch verringern. Gesucht wird eine Plattform, welche gemessen an den Kosten, aber auch an innovativer Funktionalität, Licht genauso selbstverständlich auf winzigen Chips kontrollierbar macht, wie man es von elektrischen Signalen in der Elektronik gewöhnt ist.

In diesem noch offenen Technologierennen sind besonderes jene Plattformen interessant, welche eine gewisse Universalität, d.h. ein Offenheit für verschiedene Anwendungsbereiche mitbringen. Gleichzeitig muss die Produktion in den Stückzahlen so skalierbar sein, dass die resultierenden Produkte möglichst vielen Menschen zugute kommen können. Ein Favorit in diesem Wettbewerb ist der Ansatz der Silizium-Photonik, d.h. das Bestreben, optische Strukturen mit Silizium als Grundmaterial zu realisieren. Es liegt auf der Hand, dass die bereits existierende, enorm weit entwickelte Infrastruktur der CMOS-Technologie zu diesem Zweck genutzt werden könnte. In der Tat sind bereits viele große Unternehmen mit Erfahrung in der CMOS-Technologie, wie z.B. Intel, IBM, Samsung und weitere, im Begriff, die Silizium-Photonik zur Marktreife zu entwickeln. Die Verwendung von Silicon-On-Insulator-Wafern (SOI) verspricht eine hohe Dichte optischer Komponenten. Gleichzeitig versucht man mit dieser Technologie, die Kompatibilität in der Herstellung von photonischen, integrierten Schalkreisen mit den etablierten Fabrikationsprozessen der Elektronik nach überschaubaren Anpassungen zu wahren. Dieser Ansatz wird in der SOI-Plattform verfolgt. Es erweist sich als praktisch, dass Silizium für infrarotes Licht, also auch für Wellenlängen der optischen Telekommunikation, transparent ist und einen sehr hohen Brechungsindex aufweist. Das bedeutet, dass das Licht durch Silizium-Strukturen sehr eng geführt werden kann und sich daher höchste Integrationsdichten erreichen lassen.

Auch wenn Silizium viele vorteilhafte Eigenschaften besitzt, so fehlen gewisse physikalische Effekte, die wichtige Anwendungsbereiche erschließen würden. Zu nennen ist z.B. eine optische Nichtlinearität zweiter Ordnung ($\chi^{(2)}$-Nichtlinearität). Der lineare elektro-optische (Pockels-)Effekt beispielsweise wird zur Zeit in vielen kommerziellen Modulatoren für die Langstrecken-kommunikation genutzt. Außerdem ist es schwer, einen Laser auf Silizium zu realisieren, da das Material ein indirekter Halbleiter ist, stimulierte Emission daher zusätzlicher Phononen bedarf und folglich ein sehr unwahrscheinlicher Prozess ist. Die oben genannten Unternehmen und auch die Mehrzahl von Forschungseinrichtungen untersuchen Alternativen, welche auf der Integration von geeigneten anorganischen Materialien wie III-V-Halbleiter auf Silizium basieren.

Organische Materialien stellen eine weitere Materialklasse dar, welche fehlende Fähigkeiten des Siliziums ausgleichen kann. Die Integration von organischen Materialien mit Silizium führt zu silizium-organischen Hybrid-Bauteilen (SOH) auf der sogenannten SOH-Plattform. Tatsächlich ist der Freiraum und das Potential, ein passendes, organisches Material zu finden, nur begrenzt durch den erforderlichen Materialentwicklungsaufwand, welcher die Gewährleistung der Langzeitstabilität mit einschließt. Die Integration des organischen Materials wird oftmals durch eine einfache Rotationsbeschichtung (spin coating) gelöst, welche eine organische Deckschicht auf dem Silizium-Chip mit direktem Kontakt zu den betreffenden Wellenleitersektionen liefert.

Die Evaluierung des Potentials der SOH-Plattform ist Aufgabe der vorliegenden Arbeit. Mit der Herstellung und Charakterisierung von Prototypen wird der Nachweis der Praktikabilität des SOH-Ansatzes für ausgewählte Anwendungen erbracht.

Kapitel 1 beschreibt den Stand der Technik und gibt einen knappen Überblick über gegenwärtige Plattformen für PICs.

Kapitel 2 vermittelt einen Überblick über den theoretischen und technischen Hintergrund. Dazu zählt eine Zusammenfassung der wichtigsten, relevanten Eigenschaften von Silizium.

Vier verschiedene Si-Wellenleitertypen auf einer isolierenden, dielektrischen Unterlage (SOI) werden vorgestellt: (a) Der gewöhnliche Streifenwellenleiter mit rechteckigem Profil. (b) Der Rippenwellenleiter in einer Ausführung, die besonders niedrige Propagationsverluste verspricht und daher genutzt wird, um Licht über größere Distanzen auf dem Chip zu führen. (c) Für die Herstellung aktiver Bauteile ist eine starke Interaktion des Lichts mit einem (hier: organischen) Deckmaterial gewünscht. Ein Schlitzwellenleiter (*engl.* slot waveguide) liefert diese Eigenschaft. (d) Um Schlitzwellenleiter elektrisch ansteuern zu können, werden sie über einen dünnen Streifen (Sockel) aus Silizium mit weiter entfernt liegenden Metallelektroden kontaktiert. Ein Mindestabstand zwischen Metall und optischem Wellenleiter ist unabdingbar, um eine Dämpfung des Lichtes durch das Metall zu vermeiden. Dieser Wellenleitertyp nennt sich streifen¬belasteter Wellenleiter (*engl.* strip-loaded slot waveguide, *deutsch* auch Sockel-Wellenleiter).

Die optischen Verluste in Siliziumwellenleitern stammen nur zu einem kleinen Teil aus der Absorption im Material (das gilt für intrinsisches Material oder im Fall schwacher Dotierung). Der größere Dämpfungsanteil resultiert aus Lichtstreuung an den fabrikationsbedingt rauhen Seitenwänden. In Zusammenarbeit mit der finnischen Gruppe um Seppo Honkanen wurde gezeigt, dass sich diese Streuverluste durch dünne Schichten verringern[2] lassen, die Atomlage für Atomlage aufgebracht werden (engl. atomic layer deposition, ALD) und eine hohen Brechungsindex aufweisen.

Weiter werden in Kapitel 2 zur Vorbereitung der Diskussion eines SOH-Laser-Prototypen verschiedene Farbstoffe vorgestellt, die sich zur Einbettung in organische Materialien eignen. Ferner werden der Plasmadispersionseffekt sowie der lineare elektro-optische Effekt diskutiert. Einige grundlegende Fragestellungen beim Design von SOH-Modulatoren in Wanderwellen-Konfiguration werden erläutert. Schließlich werden übliche Modulationsformate für die optische digitale Kommunikation kurz vorgestellt.

Die folgenden Kapitel 3 und 4 beschreiben die konkrete Umsetzung der vorgestellten Konzepte in Form von Prototypen, um dadurch grundsätzliche

---

[2] Im Anwendungsfall von nur mit Luft bedeckten Wellenleitern; z.B. für die Detektion von ausgewählten Molekülsorten mittels spezifischer Rezeptormoleküle am Wellenleiter.

Fragen zur Funktionalität der SOH-Plattform zu klären: Wie lässt sich Licht in SOH-Bauteilen erzeugen? Wie lassen sich mit SOH-Bauteilen elektrische Signale zu optischen Signalen konvertieren? Diese Ergebnisse werden in zwei gesonderten Abschnitten *SOH-Technologie zur Erzeugung* und *zur Modulation von Licht* am Ende dieser Zusammenfassung diskutiert.

Kapitel 5 vergleicht SOH Modulatoren mit entsprechenden Bauteilen, die auf anderen Plattformen realisiert wurden. Dabei zeigte es sich, dass Modulatoren basierend auf dem Plasmaeffekt nicht nur Datenströme bis zu 35 Gbit/s einem optischen Träger aufmodulieren können, sondern dass beim Einstrahlen eines optischen Signals mit 35 Gbit/s dasselbe Bauteil auch als photoelektrischer Wandler fungiert und wie eine Photodiode optische On-off-Signale detektieren kann. Bemerkenswert ist dabei, dass die Bandlücken-Energie von Silizium größer ist als die Energie der einfallen Photonen. Als Ursache für diesen photo-elektrischen Effekt werden Defektzustände aufgrund des Dotiervorgangs vermutet. Da die Empfindlichkeit dieses Photodetektors nicht den üblichen Anforderungen entspricht, wird er wohl nicht als Detektor in empfindlichen Empfängern verwendet werden. Allerdings könnte ein solcher Detektor als Monitor dienen, wie z.B. beim Abstimmen des Arbeitspunktes eines Mach-Zehnder-Modulators (MZM) oder eines IQ-Modulators.

Verglichen werden die SOH-Modulatoren ferner mit Modulatoren auf der GaAs-Plattform. Die Untersuchungen fanden im Rahmen einer Industriekooperation (u^2t Photonics UK) bei Datenraten bis zu 150 Gbit/s statt.

In einer kurzen Zusammenfassung werden die Hauptmerkmale der verschiedenen Plattformen aufgelistet. Nach unseren Untersuchungen haben GaAs-Modulatoren gegenüber SOH- und SOI-Modulatoren gewisse Vorteile, was Einfügedämpfung, Bandbreite und Flexibilität bei der Wahl der Modulationsformate angeht. Andererseits machen Stärken der SOH-Modulatoren bei der Fertigung mit etablierten CMOS-Prozessen ihren Einsatz in bestimmten Anwendungsgebieten wahrscheinlich. Insbesondere der wesentlich kleinere Formfaktor und das Potential für sehr geringere Treiberspannungen können helfen, den Energieverbrauch zu senken.

Kapitel 6 schließt diese Arbeit mit einem kurzen Ausblick über die künftige Entwicklung der SOH-Plattform ab. Die SOH-Technologie ermöglicht hochintegrierte photonische Schaltkreise für die Datenübertragung. Dabei können zahlreiche Komponenten parallel integriert werden und unterstützen daher auf natürliche Weise die verschiedensten Multiplexverfahren. Hierfür können Konzepte, welche sich bereits in Systemexperimenten mit diskreten Komponenten bewährt hatten, auf die Besonderheiten photonischer Schaltkreise angepasst werden. Als Beispiel sei das OFDM-Verfahren (*engl.* orthogonal frequency division multiplexing) genannt. Es ermöglicht die Nutzung von spektral besonders dicht benachbarten optischen Trägern zur Datenübertragung mit hohen aggregierten Datenraten.

Zu Kapitel 3: SOH-Technologie zur Erzeugung von Licht wird am Beispiel eines SOH Lasers vorgestellt. Siliziumstrukturen wie Streifen- oder Schlitzwellenleiter bilden die Grundlage für den Resonator. Ein Deckmaterial, welches aus einem Laserfarbstoff-Polymer-Verbund besteht, wird durch optisches Pumpen bei einer Wellenlänge von 1064 nm zu stimulierter Emission bei einer Wellenlänge von 1310 nm angeregt. Die Wiederholrate des Pump-Pulses beträgt 13.7 Hz. In der Folge wird erstmalig ein im Infraroten emittierender SOH-Laser demonstriert, der Pulsspitzenleistungen von 1 W an der Austrittsfacette des Lasers auf dem Chip aufweist. Davon können bis zu 365 mW in eine Faser eingekoppelt werden. Zum Vergleich: Die größten Emissionsleistungen anderer, auf der Integration anorganischer Materialien beruhender Laser auf Silizium liegen gegenwärtig bei weniger als 50 mW an der Laseraustrittsfläche auf dem Chip.

Die bemerkenswert hohe Spitzenleistung genügt für nichtlineare Anwendungen auf dem Siliziumchip. Der Pulsbetrieb des Lasers ist insbesondere für Anwendungen wie (möglicherweise nichtlineare) Spektros-kopie für chemische Analysen kein Nachteil. Es gibt zahlreiche weitere Laserfarbstoffe, welche zur Erzeugung anderer interessanter IR-Wellenlängen führen, wenn der optische Resonator geeignet dimensioniert wird. Auch wenn von Laserfarbstoffen allgemein nicht die gleiche Langzeitstabilität wie von Halbleiterlasern erwartet werden kann (mangelnde Photostabilität), so stellten wir in unseren Untersuchungen fest, dass der hier verwendete Farbstoff (IR 26) sich über Wochen intensiver Messungen stabil im Vergleich zu anderen Farbstoffen gezeigt hat. Außerdem ist die Herstellung dieser Laser potentiell sehr kostengünstig (von der Pumplichtquelle einmal abgesehen). Anwendungen, die die einmalige Verwendung und anschließende Entsorgung eines Analysechips[3] voraussetzen, können so um eine der wichtigsten Schlüsselkomponenten erweitert werden.

Eine attraktive, zukünftige Anwendung wäre die Herstellung von Spektrometern in ressourcenschonender (miniaturisierter) Form als Wegwerfprodukt in der Art von Blutzucker-Messstreifen. Dies könnte es jedem Haushalt ermöglichen, sich z.B. über Schadstoffe in Lebensmitteln zu informieren und dadurch eine verbesserte industrielle Produktion zu erzwingen.

Zu Kapitel 4: SOH Technologie zur Modulation von Licht wird vorgestellt. Die Wahl des organischen Deckmaterials entscheidet über mögliche Anwendungsszenarien. Es werden Modulatoren mit organischen Kristallen sowie mit Chromophoren in einer Polymermatrix konzipiert, hergestellt (teilweise mit externen Kooperationspartnern) und charakterisiert.

In dieser Arbeit wird erstmalig ein für hohe Datenraten (12.5 Gbit/s) geeigneter SOH-Mach-Zehnder Modulator demonstriert, in welchem das Deckmaterial aus der Klasse der organischen Kristalle mit einer nennenswerten $\chi^{(2)}$-Nichtlinearität stammt. Auch wenn die Modulationsempfindlichkeit soweit nur der eines durchschnittlichen Plasma-Effekt-Modulators entspricht, so zeigt dieser Versuch doch folgendes: (a) Es ist möglich, organische Kristalle auf Siliziumchips mit CMOS-artigem Aufbau zu integrieren, d.h. Hochfrequenzelektroden und ein elektro-optischer Kristall können mit stark führenden Wellenleitern kombiniert werden. (b) Die $\chi^{(2)}$-Nichtlinearität steht auch für andere Anwendungen, wie z.B. für die Erzeugung von THz-Wellen oder für parameterische Verstärkung zur Verfügung. Als weiterer Vorteil zählt, dass im Wellenleiter geführtes Licht viel höhere Intensitäten erreicht, als es in einer Freistrahlkonfiguration über längere Strecken möglich ist. Dies ist mit entsprechenden Vorteilen für die Effizienz verbunden. (c) Potentiell kann eine ganze Matrix von Hochgeschwindigkeitsmodulatoren parallel arbeiten, was für Multiplex-Anwendungen von Bedeutung ist. (d) Je nach Kristall könnte sich die Temperaturstabilität eines solchen SOH Modulators auf z.B. 250°C erhöhen. Die Langzeitstabilität ist durch die stabile Kristallstruktur ohnehin gesichert.

---

[3] Wie z.B. beim Labor-auf-dem-Chip (*engl.* lab-on-chip). Eine Reinigung dieser Chips wäre unwirtschaftlich.

Unter Verwendung von Chromophoren in einer Polymermatrix[4] wurde erstmalig ein SOH-Modulator für komplexe Modulationsformate demonstriert. Die Bitfehlerwahrscheinlichkeit (*engl.* bit-error-ratio, BER) des durch diesen IQ-Modulator generierten Datenstroms übertraf zum Zeitpunkt der Veröffentlichung die BER von nicht-resonanten Plasma-Effekt-Modulatoren. So konnte eine QPSK-Modulation bei 28 GBd, d.h. bei 56 Gbit/s, ohne Vorverzerrung oder Entzerrung gezeigt werden (bei einer Bitfehlerwahrscheinlichkeit von 4.5×10^{-4}, also unterhalb der Standardgrenze für Vorwärtsfehlerkorrektur[5]). Mit einem Entzerrer lässt sich die Vorwärtsfehlerkorrektur sogar einsparen. In einem weiteren Experiment wurden Daten bei 28 GBd im 16QAM Format generiert, das entspricht 112 Gbit/s. Eine Vorverzerrung war ausreichend, um mit der Bitfehlerwahrscheinlichkeit von 1.2×10^{-3} unterhalb der Standardgrenze für Vorwärtsfehlerkorrektur zu bleiben. Die Einfügedämpfung ist im Demonstrator relativ hoch und muss gegenwärtig mit einem optischen Verstärker ausgeglichen werden. Wegen dieses zusätzlichen Aufwands lässt der gegenwärtige Entwicklungsstand diesen SOH-Modulator eher für Langstreckenkommunikation mit kohärentem Empfang als geeignet erscheinen. Neuere nichtlineare Polymere würden auf den gleichen Si-Strukturen einen Einsatz der Modulatoren in Datenzentren interessant machen, da sich diese Modulatoren mit extrem kleinen Spannungen und somit bei geringem Energieverbrauch betreiben lassen.

Obwohl eine Integration von elektronischen und optischen Komponenten auf dem gleichen Chip Vorteile verspricht, ist die Kompatibilität mit CMOS-Prozessen noch nicht gewährleistet (ohne das Aufbringen der organischen Deckschicht zu berücksichtigen). In naher Zukunft werden der elektronische und der optische Teil des Chips wahrscheinlich auf getrennten Substraten realisiert und in hybrider Integration vereinigt. Für die zu erwartenden Stückzahlen ist eine elektrische Treiberschaltung, die in Silizium-Germanium-Technologie (SiGe) hergestellt wird, vermutlich wesentlich kostengünstiger. Außerdem kann diese Technologie größere Spannungen bei höheren Frequenzen liefern, was Sender mit größerer Datenkapazität ermöglicht. Ein SOH-

[4] Chromophore in Polymermatrix. Das hier verwendete, kommerziell erhältliche Material wird bereits in Polymermodulatoren von GigOptix Inc. eingesetzt.

[5] Vorwärtsfehlerkorrektur (*engl.* forward error correction, FEC). Mit dieser Methode lässt sich ein hinreichend fehlerarmer Kanal gewährleisten.

Modulator kann direkt von einem SiGe-Schaltkreis getrieben werden, wie im Rahmen einer Industriekooperation mit GigOptix Inc. demonstriert wurde. Der SiGe-Treiber liefert ein Signal mit einer Amplitude von 0.3 V, was zum Betrieb eines SOH-Mach-Zehnder Modulators bei 12.5 Gbit/s und geringer Bitfehlerwahrscheinlichkeit genügt.

Preface

Silicon photonics promises to duplicate the spectacular development and resulting progress from miniaturization of electronics to very dense integration. Silicon photonics is expected to bring the advantages of integration and scaling to optics.

The transition from bulky electronic circuits in highly integrated circuits (IC) has enabled today's information society. Making devices such as the personal computer and smart phones accessible to the public for interconnecting individuals, changed the way of social interactions and of conducting business. The continuation of this development can be sustained only through an ever more efficient use of resources such as energy and raw materials. If not, this information society might stop evolving. A look back into history offers the dire picture of civilizations that lose their momentum. To transport the exponentially increasing amount of data, but also to more efficiently derive more and more information (keywords: chemical/biological sensors, lab-on-chip), optical methods promise to deliver spectacular progress.

By switching to optical solutions (e.g., by using optical fibers instead of coaxial cables, or by using optical sensors) a first leap of progress was already made. Following the lead of electronics, a second giant leap in optics is imminent: Miniaturization of optical components in form of photonic integrated circuits (PIC) will help to save power. Miniaturization promises large gain in functionality and a general increase in capabilities of each PIC by massive parallelization. To make this happen, a technological platform is required, where photons can be controlled as easily and naturally on-chip as we control charges in microelectronics.

A number of platforms compete to make this vision reality. They have to prove their universality for use in many applications scenarios at a reasonable cost per piece. In addition, the scalability to mass production has to be shown, so that more than just a few people benefit from this innovation. Given the experience in CMOS technology (complementary metal oxide semiconductor) for reliable production with sufficient yield, and the vast infrastructure in place, it is a straightforward idea to attempt adapting this infrastructure for PICs. And indeed, many companies including Intel, IBM, Samsung, Luxtera, Kotura,

Lightwire and others are investing heavily to develop the so-called silicon pho-photonics platform. More precisely, special silicon-on-insulator (SOI) wafers are used to create waveguides (WG) with strong light confinement, while maintaining compatibility to existing CMOS processes with minor adaptations. Silicon is transparent at IR, especially at telecommunications wavelengths. In addition, silicon features a large refractive index enabling very dense optical integration. Silicon photonics offers the prospect of integrating optical functionality right next to electronics on the same chip.

In spite of the many advantages of Si, some currently popular physical effects are missing. For instance, the linear electro-optical effect is used almost exclusively for modulation in long-haul communications today. Moreover, it is notoriously hard to achieve lasing on silicon due to its indirect bandgap. The companies mentioned above and many research institutions investigate the hybrid integration of other, inorganic materials on silicon to make a laser, and they contend with the plasma dispersion effect of silicon for modulators. This platform is referred to as SOI platform.

In this thesis we explore a very similar, scalable, yet distinctively different platform. Instead of using inorganic materials for a hybrid integration technique, we choose from the vast range of organic materials. These can be engineered to exhibit the desired properties. This so-called silicon-organic hybrid (SOH) platform combines organic materials with silicon photonic structures.

To evaluate the potential of the SOH platform, prototypes of key components (SOH lasers, SOH modulators) are conceptualized, designed, fabricated (also "fabless"), post-processed and characterized. Only the part of post-processing and the subsequent steps differ from the usual SOI platform. An organic cladding is deposited with a proven method, e.g., by spin-coating as used throughout most CMOS processes. The cladding is left to cover the entire wafer. The SOH devices' specific advantages and shortcomings are identified. With regard to the corresponding applications, the added value of organic claddings is demonstrated.

Chapter 1 gives an overview of the state of the art, including a summary of selected, currently available platforms for integrated optics.

Chapter 2 summarizes the theoretical and technological background, especially with respect to CMOS compatibility. The relevant properties of Si are listed, useful waveguide types are introduced, and loss and light interaction characteristics are discussed. The gain medium for the SOH laser, the linear electro-optic materials used to make SOH modulators and the employed modulation formats are explained.

Chapter 3 focuses on the realization of an SOH laser based on dye molecules as a gain medium.

Chapter 4 presents the results about achieving electro-optic (EO) modulation with the linear electro-optic effect from two different classes of materials: organic crystals and chromophores hosted in a polymer matrix. Successful integration and advanced modulation capability is demonstrated. A platform for making modulators would be incomplete without a concept of how to integrate the necessary electric driver circuits. The long-term goal is to use integrated CMOS drivers, probably on the same chip as the photonic structures. However, a short-term solution seems to demand a currently more cost-efficient, lower volume production (as long as consumer products are not specifically targeted). Within an industry cooperation we identified drivers made with the SiGe platform to be ideally suited to the SOH devices. This suitability is demonstrated.

Chapter 5 shows benchmarking measurements of modulators fabricated on the SOI platform (using the plasma dispersion effect), and on the GaAs platform. A short comparison of the platforms is given, pointing out the use of SOH modulators for very low energy consumption applications, for long-haul communication, and for frequency comb line generation.

Chapter 6 concludes this work with a short outlook on using the specific advantages of the SOH platform.

Achievements of the Present Work

In this thesis we study the potential of the silicon-organic hybrid (SOH) platform for integrated optics. The unique properties of selected organic materials are added to the basic silicon devices. We investigate the feasibility of this approach by making prototypes of key components in form of photonic integrated circuits (PIC): SOH lasers and SOH modulators are designed, fabricated (partly together with external partners), post-processed, and characterized. Application scenarios are identified and demonstrated in proof-of-principle experiments. A concise overview of the main achievements follows.

Technology Platform Development

Waveguide loss reduction

The insertion loss of photonic structures is important. In cooperation with the Finnish research group of *Honkanen et al.*, we investigated loss reduction for waveguides (WG) by using atomic layer deposition (ALD) to smoothen the rough sidewalls of the WGs, which cause strong light scattering. Coating silicon strip and slot WGs with a 50 nm amorphous titanium dioxide (TiO2) film reduces losses down to (2 ± 1) dB/cm and (7 ± 2) dB/cm, respectively, at a wavelength of 1.55 μm.

CMOS metal stack for SOH applications

On the one hand, SOH devices rely on the strong interaction of light in the organic material with the light guided by an optical WG. On the other hand, a metal stack similar to the one used in CMOS technology is necessary to make electrical lines in a safe distance above the optical WG layer for avoiding additional light attenuation. Hence also the optical WGs are covered with layers made of glass, silicon nitride, or silicon carbide. In cooperation with *Bogaerts et al. from IMEC in Belgium*, a procedure for the opening the metal stack layers down to the optical WGs has been developed for depositing the organic material. The resulting prototypes of high-speed SOH modulators were characterized thereby demonstrating the feasibility of the process.

Integration of organic crystals on silicon

Organic crystals possess formidable properties, but their integration on silicon is a challenge, especially on a topography as uneven as a CMOS metal stack with trenches down to the optical WGs. In cooperation with *Jazbinsek et al. from Rainbow Photonics in Switzerland*, numerous deposition methods have been tested. The resulting prototypes were characterized and a suitable method for growth was selected. Successful deposition was proven by using the linear electro-optic effect of the crystals to show electro-optic modulation. While Rainbow Photonics is currently using bulk organic crystals for THz wave generation, the integration of the same crystals on silicon promises more efficient emitters: Within an SOH WG the pump light can be spatially confined to smaller regions for achieving much higher intensities at lower power.

Demonstrations

SOH Laser

For the first time an SOH laser has been demonstrated. The device works at room temperature, is optically pumped, uses the dye IR 26 in a polymer matrix as a gain medium, and emits at a wavelength of 1310 nm. The emission pulse peak power of 1 W at the output facet of the WG is one order of magnitude larger than reported for other lasers on silicon.

High-speed organic crystal modulator

For the first time a high-speed modulator using organic crystals on a CMOS metal stack was demonstrated at 12.5 Gbit/s. This result followed from the development of organic crystal deposition for the SOH platform,

SOH IQ modulator

For the first time an SOH modulator for complex modulation formats was demonstrated at 112 Gbit/s using the 16QAM format. This makes it the modulator with the highest data rate for single-polarization and single carrier transmission with a bit error ratio (BER) below the hard decision (HD) forward error correction (FEC) limit. The measured BER is less than 1.2×10^{-3}.

SiGe driver for SOH modulator operation

The SOH modulators require a driver for supplying an electrical signal. To reduce the energy consumption of the system, external electrical amplifiers have to be avoided. The compatibility of SiGe drivers (which can be produced more economically at lower volume than CMOS drivers) with SOH MZMs has been proven experimentally. The utilized SiGe chip accepts electronic input signals with powers as low as –20 dBm at 12.5 Gbit/s, and generates an output signal with 0.3 V amplitude at 50 Ω (0 dBm). This suffices to drive an SOH modulator with on-off-keying (OOK) to generate an optical signal with a quality factor Q^2 corresponding to 11.7 dB. Here, the SiGe chip acts as a signal regenerator and signal amplifier only, but additional logical functions could be implemented as well.

Benchmarking with GaAs modulators

The evaluation of the SOH platform requires a minimum of familiarization with its closest competitors. In cooperation with *O'Keefe from u^2t Photonics UK* and others, we characterized a GaAs IQ modulator prototype at 25 GBd. Using the 32QAM format at 125 Gbit/s, a BER below the HD limit for FEC was achieved. For a 64QAM format a data rate of 150 Gbit/s was realized with a BER below the software decision (SD) FEC limit. This is the first demonstration of an IQ modulator for advanced modulations formats, which has been fabricated on the GaAs platform.

Benchmarking with SOI plasma-effect modulators

In cooperation with *Yu et al. from Ghent University, Belgium*, we measured a plasma-effect Mach-Zehnder modulator (MZM, free-carrier plasma dispersion effect) at 35 Gbit/s using on-off keying (OOK) with a BER below the HD FEC threshold.

Photonic Integrated Circuit Designs

Silicon photonic chips are the workhorse for all research in silicon photonics, e.g., research on applied physical effects, on new modulation formats, and on demonstrations of new cladding materials leading to ultra-low energy consumption of modulators or switches. The chip's design (creation of mask layout) presented a tangible contribution to our group's success. The following masks for deep UV lithography have been designed:

Mask E-DESIGN

For the purpose of providing test samples for material absorption, and for the study of waveguide losses related to scattering dependent on waveguide width.

Mask SOFI 2

For the purpose of using a CMOS metal stack for making electrodes for SOH phase modulators (PM), MZMs, and IQ modulators [J2], [J4], [J7], [C1], [C2], [C5], [C6] for optical filter structures implemented as delay interferometers, for waveguides designed to make optical parametric amplifiers with quasi-phase matching, for liquid-crystal phase shifter test structures, and for packaging investigations.

Mask SOFI 2.5

For the purpose of making packaging tests, for validating arrayed waveguide gratings (AWG) for advanced filtering (designs provided by IMEC), for photonic wire-bonding test structures, for SOH lasers, and for ultra-short liquid-crystal phase shifters.

Mask SOFI 3

For the purpose of testing SOH PMs, MZMs, IQ modulators, SOH comb line generators, integrated OFDM SOH transmitters, OFDM receiver filter structures, for high-speed packet switching, for liquid-crystal phase shifter test structures, and for packaging tests.

Mask MISTRAL-NOGATE I and III, MISTRAL-TOPGATE

For the purpose of testing SOH PMs, MZMs and IQ modulators from e-beam-based lithography. [J5], [J7]

1 Introduction – Integration Increases Performance

Photonic integration promises more efficient devices. With similar or even more elaborate functionality as discrete optical components, integrated devices are expected to consume less energy, to require much less space, and to need smaller amounts of raw materials for production. Common examples include integrated lasers, arrayed waveguide gratings (AWG) as optical filters, or entire spectrometers as a lab-on-a-chip which consume less analyte and processing chemicals.

Photonic scaling means that a large number of components with different or identical functionality can be integrated and interconnected on the same chip. Integrated optics will follow the path of electronics. The transistor radio was already introduced in 1954. Going beyond single transistors by employing integrated circuits (IC) enabled the revolution in computing. The development from Intel's processor 4004 containing 2300 transistors in 1971, to today's processors and graphic cards with more than 1 billion transistors serves as a development template from discrete optics to integrated nano-photonic[6] ICs.

1.1 Available Platforms for Photonic Integration

Discrete optical components are often made of glasses or materials chosen for a very specific purpose, such as $LiNbO_3$ for modulators, III-V semiconductors for lasers, liquid crystals with some substrate for displays, or metals as coatings for gratings in filters. Integrating many hard-to-process materials on the same chip is a challenge. In the following, a number of platforms are discussed:

The silica-on-silicon platform

WGs are made of SiO_2, which only provides a low refractive index contrast[7] of $\Delta n = 0.01...0.1$. That allows for a mode diameter of around 8 μm for the guided light, and hence imposes a lower limit of bend radii to around 5 mm. While this platform offers low optical loss and easy packaging, devices cannot be integrated very densely. Furthermore it takes significant effort to integrate

---

[6] Optical waveguides with sub-micrometer width.

[7] Waveguiding by total internal reflection needs a higher refractive index for the core than for the surrounding cladding.

functionality beyond simple waveguiding. This is usually accomplished by manually inserting discrete components made of other materials, e.g., $LiNbO_3$ or polymers.

III-V compound semiconductors

Compounds such as GaAs, InP, offer a range of very interesting properties, as the bandgap and conductivity can be easily engineered.

The indium phosphide (InP) platform [3] shows an index contrast of $\Delta n = 0.2...0.5$ for optical WGs. The mode diameter of around 2 µm implies a lower limit of the bend radius to 0.5 mm. Currently, this platform successfully entered the long-haul market, where dense wavelength division multiplexing (DWDM) is of interest, in spite of small wafer sizes (2…4”). Yield and integrated functionality from lasers, high-speed modulators [4], AWGs and high-speed detectors make a very convincing business case.

The gallium arsenide (GaAs) platform employs larger wafers (6”) than the InP platform. A mature infrastructure for making power amplifiers in cell phones and monolithic microwave integrated circuits (MMIC) is in place. However, there are no lasers or detectors commercially available in GaAs at 1550 nm, and the typical modulator length of 3 cm is comparatively large.

Silicon-photonic circuits

Silicon-photonic circuits are in most cases intended for CMOS compatible fabrication, in order to make use of the CMOS industry’s large infrastructure and expertise.

The silicon-on-insulator (SOI) platform features a very large index contrast of $\Delta n = 1.0 \ldots 2.5$ for typical claddings. The mode diameter can be around 0.4 µm which places the lower limit for bends at radii around just 5 µm. Also modulators [5] can be made short with a length of 0.5 …3 mm (in non-resonant configuration). To make lasers or detectors, additional materials are integrated. The hybrid integration of inorganic materials is often referred to as being part of the SOI platform. For instance, integrating germanium (approach of Intel, IBM, and many others) shows promise to serve for high-speed detectors [6] and also to make lasers [7] on silicon.

The silicon-organic hybrid (SOH) platform is based on the SOI platform, but integrates missing functionalities of silicon by incorporating organic materials [C12], [8]. Given a virtually unlimited choice of organic materials and engineering possibilities, the unique properties of materials of the organic world of chemistry can be added to these silicon-based PICs. *This is the platform under investigation in this thesis.*

Silicon-on-sapphire (SOS) is another modification of the SOI platform, but avoids buried silicon oxide as insulator. Instead, sapphire is used to obtain a good transparency even for longer wavelengths than are usable with SOI, making it a very interesting candidate for mid-IR applications.

In spite of the challenge to make lasers or detectors for IR light on silicon, a business case is forming for the silicon platform. It uses much larger wafers than all other platforms (diameter 8 in…12 in). This is important, because hundreds of processing steps need to be performed for each wafer. Dense integration allows devices with small footprint (10 000 times smaller as compared to the silica-on-silicon platform). Provided that integration of additional functionality succeeds, the silicon platform is expected to offer a cost advantage for large production volumes when compared to the other platforms. The silicon platform might also be the only scalable platform to make affordable consumer products.

1.2 Employing CMOS-Based Silicon Photonic Processes

Adhering to CMOS process standards achieves compatibility with the CMOS infrastructure, and this comes with additional benefits. A lesson to be learned from the integrated electronics industry is that the risks related to fabrication can be minimized by using the foundry model. Activities like design, packaging, quality control, and marketing are separated from the fabrication, which is outsourced to a foundry[8] – another company which focuses solely on fabrication for multiple clients. Thus, the foundry can flexibly react to market demands.

For those interested to make highly complex and powerful PICs, additional advantages can be found: A project manager will value the volume scalability and reliability of production from a foundry. This considerably reduces risks

[8] Examples of CMOS foundries: GlobalFoundries, freescale, tsmc (Taiwan semiconductor manufacturing company). For research purposes: imec in Belgium, CEA Leti in France, IME in Singapore.

and cost compared to an in-house production. A product engineer can rely on the extensive design specifications a foundry has to supply, and he can count on sufficient fabrication yield. In a research setting, a PhD candidate can make a PIC design and concentrate on the scientific aspects, while an experienced fab-fabrication facility[8] takes care of the routine work of structuring the SOI wafers.

For silicon photonic processes, only a large commercial demand would enforce common standards that are required before any large CMOS foundry will start implementing the inevitable adaptations of its processes to make PICs.

The fact that significant industry interest has already formed around the processing of silicon for photonic purposes leads to a preference of materials. These materials include:

- Crystalline silicon
  (c-Si, provided as wafer, adjustable electronic properties),
- Poly-crystalline silicon
  (poly-Si, from deposition, adjustable electronic properties),
- Amorphous silicon
  (a-Si, from deposition, limited choice of electronic properties, unstable).
- Silicon nitride
  (Si_3N_4, from deposition, limited choice of electronic properties, excellent low loss for light propagation [9]),

Each of these materials and their combinations can be used for making photonic integrated circuits on silicon substrates. Multiple layers could be deposited and allow making even more complex circuits, i.e., a 3D silicon photonic stack (multiple layers, each with optical waveguides). In this work we limit our study to crystalline silicon. For benchmarking also GaAs is investigated.

1.3 Light Generation Using the SOI Platform

An on-chip light source could contribute to or enable a number of applications, such as data communication for optical interconnects, or even intra-chip communication (e.g., clock signal distribution). In addition, some biomedical or sensing applications (e.g., measuring the concentration of an analyte) also require a light source. In short, on-chip sources are obviously of use for any op-optically integrated device, especially to reduce coupling efforts[9] of light.

An overview of lasing on silicon is given by Liang and Bowers [10]. We concentrate here on laser emission at IR wavelengths, although there are other interesting ways to obtain light emission, e.g., by third harmonic generation using photonic crystals [11]. Also, light emitting diodes using photonic crystals of silicon, which are electrically pumped, have been fabricated and obtain low-power (4 pW) narrow emission lines at a wavelength of 1515 nm at room temperature [12]. Optically active defects were introduced in Si by treating it with hydrogen plasma.

To make a silicon laser, several different approaches can be distinguished:

Raman lasers

Silicon lasers using the Raman effect need an optical pump. A recent overview is given by R. Baets in [13]. The first continuous wave (cw) Raman laser [14] still required hundreds of mW of pump light for laser emission of up to 10 mW. A lower threshold of 20 mW was shown in [15] for lasing from a ring cavity. This laser can deliver up to 50 mW of output power.

Recently, a continuous wave (cw) micrometer-scale Raman silicon laser (less than 10 μm × 2 μm footprint) based on a photonic crystal cavity an having a microwatt threshold was demonstrated [16]. At a wavelength of 1428 nm the pump light generates laser light at 1543 nm.

Given the low pump threshold, multiples of these lasers could be optically pumped on-chip from a single external laser, which would be distributed exactly where needed on-chip [13].

Epitaxially grown lasers on silicon

To use a material with a more favorable bandgap than Si, other suitable lasing materials could be epitaxially grown directly on silicon. However, the lattice mismatch between Si and semiconductors such as Ge and InP disturbs the

[9] For instance pump light can be coupled once to the chip to optically pump an array of on-chip lasers made for different wavelengths. Otherwise providing light with multiple, but separated wavelengths would require coupling light of each wavelength separately or using optical filters for multiplexing and de-multiplexing.

crystalline growth. Michel et al. overcame this issue and realized the first opti- optically pumped germanium-on-silicon laser [17]. Later on, they built an electrically pumped Ge laser [18]. The latter device emits at room temperature a multimoded beam with 1 mW of output power. The gain spectrum is rather wide (up to 200 nm), and laser lines at 1576 nm, 1622 nm and 1656 nm have been demonstrated. This type of silicon laser has a very small footprint and shows promise for applications ranging from short distance chip-to-chip communication to even intra-chip communication. Note that Ge integrated on SOI by the same research group has also been used to make high-speed photo detectors.

Hybrid silicon lasers

The InP platform allows fabricating efficient lasers with high yield. Flip-chip bonding of already validated InP laser diodes directly on silicon wafers is a most cumbersome process due to alignment and yield issues.

An unstructured InP stack can be bonded to the silicon wafer before it is structured. This can be done wafer-to-wafer, or because of yield issues die-to-wafer, using molecular bonding, or adhesive bonding. In demonstrated prototypes the resonator is realized in the silicon layer, while the gain is contributed by the InP stack, either by evanescent coupling or by actually guiding the light in the InP stack. Fang et al. demonstrated a distributed feedback silicon evanescent laser [19]. The highest reported output power of such a III-V-Si hybrid laser is 29 mW [20]. This type of laser is mostly advertised for data communication applications.

In this thesis we explore the potential of using organic materials, dye molecules in particular, to create a high peak power laser using the silicon-organic hybrid (SOH) approach. Especially applications in spectroscopy and sensing can make use of a high light intensity, either to exploit a nonlinear property, e.g., of an analyte, or simply to allow measurements of strongly absorbing substances. Investigating lasing from dye molecules can also be seen as an intermediate step, prior to testing other gain materials including quantum dots (with the potential for electrical pumping) and erbium-doped materials in similar silicon structures.

1.4 Data Communication Using the SOI Platform

The information society relies on the further development of the internet. There is a demand for: 1. Continued exponential growth [1] of transportation capacity. 2. The continued growth of data centers for services, such as enabled by cloud computing.

To demand 1: A further increase of capacity is required in networks, which constitute [21] the internet at various scales:

(a) The optical wide area network (WAN, which bridges distances of hundreds to thousands of kilometers including the global backbone, and other long-haul networks) is currently in a transition to higher data rates from 40 Gbit/s to 100 Gbit/s per optical subcarrier. While symbol rates are limited by the electronic driving circuits, the bitrates increase due to the choice of advanced modulation formats. In addition, multiplexing techniques, such as dense wavelength division multiplexing (DWDM), are used. A transition to even higher spectral efficiency seems inevitable. Nyquist-WDM or OFDM (orthogonal frequency division multiplexing) can deliver higher spectral efficiencies [22]. This case demands for PICs with integrated modulators that support advanced modulation formats, filter structures such as arrayed waveguide gratings (AWG), and delay interferometer cascades.

(b) Also the capacity of the optical metropolitan-area network (MAN) has to grow. This includes the metro core (typical circumference of 100…200 km), and metro access networks (typical circumference of 30…60 km). These are often implemented as ring networks, using DWDM or coarse WDM, although with fewer channels than in a WAN.

(c) The access network, i.e., the network connecting the subscriber to the metro access network, still uses twisted-pair copper lines in many cases (e.g., DSL in Germany). Most recently (August 2013), Alcatel-Lucent demonstrated the current limits for such copper lines [23]: Using the G.fast transmission standard, and a special technique called vectoring for cancelling noise from crosstalk, a data rate of 1.1 Gbit/s can be achieved over 70 m of good quality cable, and 0.5 Gbit/s over 100 m for unshielded cable. To generate the noise cancelation signals, this technology comes at the cost of increased demands for digital processing which in turn increases energy consumption. The alternative is to connect subscribers directly with optical lines, which is called fiber-to-the-home (FTTH). This can be realized for example as a passive optical network (PON). Access networks require large piece numbers while placing challenging demands on cost and thus present another use case for inte-grated optics.

To demand 2, continued internet growth also depends on its data centers: These already take dimensions larger than warehouses and are preferably placed in regions with cooler climate and cheap energy supply. The communication between servers in such a data center requires data transport over tens to hundreds of meters at highest data rates. Optical methods are well suited for this task, but the optical interconnects (server-to-server connections) have to be particularly small and energy efficient because of space and cost constraints [2]. This makes another use case for PICs.

In this thesis we concentrate on one key component to address the demands mentioned above, namely on silicon modulators in a non-resonant configuration, to later allow their use independently of wavelength. Current modulators fabricated on the SOI platform reach data rates up to 50 Gbit/s [24] using on-off keying (OOK, a modulation format for direct detection, see 0). SOH modulators may compete [25] with these devices in terms of bandwidth.

Using higher order modulation formats, the data rate can be increased to 112 Gbit/s, as shown by P. Dong using an SOI-modulator. This data rate can be even doubled, when using polarization multiplexing [26]. The SOH modulator demonstration at 112 Gbit/s as reported in Section 4.2 uses an organic cladding of chromophores hosted in a polymer matrix and features superior modulation sensitivity, compared to the device of P. Dong, resulting in better performance in terms of the resulting bit-error ratio (BER). The targeted application scenario for this modulator targets are long-haul networks.

Another type of SOH modulator is discussed in section 4.1 and uses organic crystals as a cladding. This is a proof-of-principle experiment for the employed material class. However, showing high-speed modulation of 12.5 Gbit/s proves that the material is integrated well enough to enable a range of other application otherwise not available on the SOI platform.

2 Theoretical and Technological Background

This chapter covers the theoretical background for this thesis, namely: The propagation of electromagnetic waves, and how to harness control of these waves to enable or to avoid their interaction by properly shaping SOI waveguides from CMOS-based fabrication technology. A specific example for such an interaction is discussed for stimulated light emission with SOH structures, and another example is presented which employs the interaction of an optical and a radio-frequency (RF) wave to obtain modulation of light. An overview about the used modulation and multiplexing techniques is given. Many design decisions and challenges of this work are driven by technological constraints.

2.1 Fundamentals

The propagation of electromagnetic (EM) waves has been extensively studied for more than a century. In spite of the complexity of real devices due to their geometry and number of (even nonlinear) materials involved, the origin of the behavior of EM fields propagating in these devices can be partly understood analytically. To introduce the symbols used in this work, a number of fundamental relations are restated here. Vectors and matrices are printed in bold letters.

Time t and the spatial coordinate $\mathbf{r} = x\mathbf{e}_x + y\mathbf{e}_y + z\mathbf{e}_z$ (unit vectors $\mathbf{e}_\mathrm{i}$ spanning a Cartesian coordinate system) are used to describe the following quantities: The electric field vector $\boldsymbol{\mathcal{E}}(\mathbf{r},t)$, the magnetic field vector $\boldsymbol{\mathcal{H}}(\mathbf{r},t)$, the electric flux density $\boldsymbol{\mathcal{D}}(\mathbf{r},t)$, the magnetic flux density $\boldsymbol{\mathcal{B}}(\mathbf{r},t)$, the electric current density $\boldsymbol{\mathcal{J}}(\mathbf{r},t)$, the electric charge density $\rho(\mathbf{r},t)$ are all real-valued. The permittivity of vacuum ε_0 and the permeability of vacuum μ_0 are constants and defined in Appendix C.1.

2.1.1 Maxwell's Equations

Maxwell's equations (p. 6 in [27]) can be written as

$$\nabla \times \boldsymbol{\mathcal{E}}(\mathbf{r},t) = -\frac{\partial \boldsymbol{\mathcal{B}}(\mathbf{r},t)}{\partial t}, \tag{2.1}$$

$$\nabla \times \boldsymbol{\mathcal{H}}(\mathbf{r},t) = \boldsymbol{\mathcal{J}}(\mathbf{r},t) + \frac{\partial \boldsymbol{\mathcal{D}}(\mathbf{r},t)}{\partial t}, \tag{2.2}$$

$$\nabla \cdot \boldsymbol{\mathcal{D}}(\mathbf{r},t) = \rho(\mathbf{r}), \tag{2.3}$$

$$\nabla \cdot \boldsymbol{\mathcal{B}}(\mathbf{r},t) = 0. \tag{2.4}$$

We assume in this subsection that there are no free charges, no free currents, and consider a nonmagnetic material. It follows that

$$\rho(\mathbf{r}) = 0 \;, \quad \boldsymbol{\mathcal{J}}(\mathbf{r},t) = 0 \;, \quad \boldsymbol{\mathcal{B}}(\mathbf{r},t) = \mu_0 \boldsymbol{\mathcal{H}}(\mathbf{r},t). \tag{2.5}$$

The real valued polarization density (in short polarization) $\boldsymbol{\mathcal{P}}(\mathbf{r},t)$ is implicitly defined using

$$\boldsymbol{\mathcal{D}}(\mathbf{r},t) = \varepsilon_0 \boldsymbol{\mathcal{E}}(\mathbf{r},t) + \boldsymbol{\mathcal{P}}(\mathbf{r},t). \tag{2.6}$$

Following p. 71 in [28] one can take the curl of Eq. (2.1), uses Eq.(2.2) and Eq. (2.6), (c is the velocity of light in vacuum, $c^{-2} = \varepsilon_0 \mu_0$) and can derive the most general form of the wave equation in nonlinear optics

$$\nabla \times \nabla \times \boldsymbol{\mathcal{E}}(\mathbf{r},t) + \frac{1}{c^2}\frac{\partial^2}{\partial t^2}\boldsymbol{\mathcal{E}}(\mathbf{r},t) = -\frac{1}{\varepsilon_0 c^2}\frac{\partial^2}{\partial t^2}\boldsymbol{\mathcal{P}}(\mathbf{r},t). \tag{2.7}$$

Assuming additionally that the medium is isotropic and homogenous, $\nabla(\nabla \cdot \mathbf{E}) = 0$, the wave equation for the electric field reads

$$\nabla^2 \boldsymbol{\mathcal{E}}(\mathbf{r},t) - \frac{1}{c^2}\frac{\partial^2}{\partial t^2}\boldsymbol{\mathcal{E}}(\mathbf{r},t) = \frac{1}{\varepsilon_0 c^2}\frac{\partial^2}{\partial t^2}\boldsymbol{\mathcal{P}}(\mathbf{r},t). \tag{2.8}$$

It is possible to split $\boldsymbol{\mathcal{P}}(\mathbf{r},t)$ into a linear and a nonlinear part,

$$\boldsymbol{\mathcal{P}}(\mathbf{r},t) = \boldsymbol{\mathcal{P}}^{(1)}(\mathbf{r},t) + \boldsymbol{\mathcal{P}}^{(\mathrm{NL})}(\mathbf{r},t). \tag{2.9}$$

2.1.2 Linear Polarization

Let us assume for the moment that the material is also lossless, and that its polarization is entirely linear. Then Eq. (2.6) can be written using the material dependent dielectric permittivity ε_r as

$$\mathcal{D}(\mathbf{r},t) = \varepsilon_0 \varepsilon_r \mathcal{E}(\mathbf{r},t) . \tag{2.10}$$

In this case, Eq. (2.8) can be written in a simpler form in Fourier space (Helmholtz equation, for the Fourier transform see Appendix A.2) using the angular frequency ω:

$$\nabla^2 \tilde{\mathcal{E}}(\mathbf{r},\omega) + \omega^2 \mu_0 \varepsilon_0 \varepsilon_r \tilde{\mathcal{E}}(\mathbf{r},\omega) = 0 . \tag{2.11}$$

The solutions of this equation are plane waves, which can be found with a separation ansatz [29] in the time domain. In this ansatz, the rapidly changing spatial (in propagation direction z) and temporal (dependent on t) contributions are written separately from an amplitude factor $A(z,t)$ (which is thus assumed to be only slowly varying) and a factor to describe the transverse field distribution. Here, we consider a group of plane waves with propagation in z direction having a central frequency ω_c. The normalized (such that the guided wave power will be $|A(z,t)|^2/2$) electric, respectively magnetic field distributions in the transverse (to the propagation direction) plane depend on the waveguide's cross section and are denoted with

$$\mathbf{E}_\mathrm{t}\left(x,y,\omega_\mathrm{c}\right) \text{ and } \mathbf{H}_\mathrm{t}\left(x,y,\omega_\mathrm{c}\right). \tag{2.12}$$

The propagation constant β containing the effective refractive index n_eff is

$$\beta = n_\mathrm{eff} \frac{\omega_\mathrm{c}}{c} \tag{2.13}$$

The solutions to the Helmholtz equation for linearly polarized light are made of analytic functions of the form

$$\mathbf{E}(\mathbf{r},t) = \underbrace{A(z,t)}_{\substack{\text{Amplitude factor,}\\ \text{slowly varying in}\\ \text{time}}} \underbrace{\exp(\mathrm{j}\omega_\mathrm{c} t)}_{\substack{\text{Rapidly varying}\\ \text{in time}}} \underbrace{\mathbf{E}_\mathrm{t}\left(x,y,\omega_\mathrm{c}\right)}_{\substack{\text{Transverse (index t)}\\ \text{distribution}\\ \text{of electric field}}} \underbrace{\exp(-\mathrm{j}\beta z)}_{\substack{\text{Propagator term,}\\ \text{rapidly varying}\\ \text{along } z}}, \tag{2.14}$$

$$\mathbf{H}(\mathbf{r},t) = \underbrace{A(z,t)}_{\substack{\text{Amplitude factor,}\\ \text{slowly varying in}\\ \text{time}}} \underbrace{\exp(\mathrm{j}\omega_\mathrm{c} t)}_{\substack{\text{Rapidly varying}\\ \text{in time}}} \underbrace{\mathbf{H}_\mathrm{t}\left(x,y,\omega_\mathrm{c}\right)}_{\substack{\text{Transverse (index t)}\\ \text{distribution}\\ \text{of magnetic field}}} \underbrace{\exp(-\mathrm{j}\beta z)}_{\substack{\text{Propagator term,}\\ \text{rapidly varying}\\ \text{along } z}}. \tag{2.15}$$

Applying the Fourier transform (see Appendix A.2 for definition) to the slowly varying envelope and the other time-dependent term gives

$$\mathrm{F}\{A(z,t)\exp(\mathrm{j}\omega_\mathrm{c}t)\} = \tilde{A}(z,\omega-\omega_\mathrm{c}). \tag{2.16}$$

The transverse functions are assumed to depend neither on t nor ω,

$$\begin{aligned}\mathbf{E}_\mathrm{t}(x,y,\omega_\mathrm{c}) &= \tilde{\mathbf{E}}_\mathrm{t}(x,y,\omega_\mathrm{c}) \\ \mathbf{H}_\mathrm{t}(x,y,\omega_\mathrm{c}) &= \tilde{\mathbf{H}}_\mathrm{t}(x,y,\omega_\mathrm{c})\end{aligned}. \tag{2.17}$$

The resulting solution functions are complex and written as

$$\tilde{\mathbf{E}}(\mathbf{r},\omega-\omega_\mathrm{c}) = \tilde{A}(z,\omega-\omega_\mathrm{c})\tilde{\mathbf{E}}_\mathrm{t}(x,y,\omega_\mathrm{c})\exp(-\mathrm{j}\beta z), \tag{2.18}$$

$$\tilde{\mathbf{H}}(\mathbf{r},\omega-\omega_\mathrm{c}) = \tilde{A}(z,\omega-\omega_\mathrm{c})\tilde{\mathbf{H}}_\mathrm{t}(x,y,\omega_\mathrm{c})\exp(-\mathrm{j}\beta z). \tag{2.19}$$

The real valued physical quantities $\mathcal{E}(\mathbf{r},t)$ and $\mathcal{H}(\mathbf{r},t)$ can be found by just taking the real part (c.c. denotes the complex conjugate) of the found solutions

$$\mathcal{E}(\mathbf{r},t) = \frac{1}{2}\left(A(z,t)\mathbf{E}_\mathrm{t}(x,y,\omega_\mathrm{c})\exp(\mathrm{j}\omega_\mathrm{c}t-\mathrm{j}\beta z)+\text{c.c.}\right), \tag{2.20}$$

$$\mathcal{H}(\mathbf{r},t) = \frac{1}{2}\left(A(z,t)\mathbf{H}_\mathrm{t}(x,y,\omega_\mathrm{c})\exp(\mathrm{j}\omega_\mathrm{c}t-\mathrm{j}\beta z)+\text{c.c.}\right). \tag{2.21}$$

The wave impedance Z_N of the medium is the ratio of the electric and magnetic field, and defined as

$$Z_\mathrm{N} = \sqrt{\frac{\mu_0}{\varepsilon_0\varepsilon_r}}. \tag{2.22}$$

2.1.3 Nonlinear Polarization

Returning to the general case which includes nonlinear polarization, it is convenient to limit all following considerations to assuming that the electric and magnetic fields can be written as a superposition of narrowband analytic functions centered at angular frequencies $\omega_1,\omega_2,\ldots$ in Eqs. (2.14) and (2.15),

$$\tilde{\mathbf{E}}(\mathbf{r},\omega) = \tilde{\mathbf{E}}_1(\mathbf{r},\omega-\omega_1)+\tilde{\mathbf{E}}_2(\mathbf{r},\omega-\omega_2)+\ldots \tag{2.23}$$

Looking at the polarization in Fourier space allows a more intuitive understanding of the nonlinear interaction between the light and medium. The polarization in the transverse plane is also written as $\mathbf{P}_\mathrm{t}(\mathbf{r},\omega_\mathrm{c})$ from here on. It depends on the susceptibility of the medium, which is usually complex to also describe the optical loss. We assume from here on (in this section) that all relevant optical frequencies are far from any resonance in the medium. For linear, isotropic media, using the Fourier transform (see Appendix A.2 for definition), the linear polarization, see Eqs. (2.9) and (2.10), can be written for the transverse components as

$$\tilde{\mathbf{P}}_{\mathrm{t}}^{(1)}(x,y,\omega_{\mathrm{c}})=\varepsilon_0\chi^{(1)}\tilde{\mathbf{E}}_{\mathrm{t}}(x,y,\omega_{\mathrm{c}})\,. \tag{2.24}$$

The second-order term of the polarization, which describes the linear electro-optic effect (also called Pockels effect), can be written [28] with $i,j,k\in\{x,y\}$ as

$$\tilde{P}_{i,\mathrm{t}}^{(2)}(x,y,\omega_{\mathrm{c}})=2\varepsilon_0\sum_{j,k}\chi_{ijk}^{(2)}(\omega_{\mathrm{c}}:0,\omega_{\mathrm{c}})\tilde{E}_{k,\mathrm{t}}(x,y,0)\tilde{E}_{j,\mathrm{t}}(x,y,\omega_{\mathrm{c}})\,. \tag{2.25}$$

This expression for the polarization $\tilde{\mathbf{P}}_{\mathrm{t}}^{(2)}(x,y,\omega_{\mathrm{c}})$ allows the quantification of the interaction of a static electric field $\tilde{\mathbf{E}}_{\mathrm{t}}(x,y,0)$ and a light field $\tilde{\mathbf{E}}_{\mathrm{t}}(x,y,\omega_{\mathrm{c}})$ existing in the same medium with the nonlinear susceptibility $\chi_{ijk}^{(2)}(\omega_{\mathrm{c}}:0,\omega_{\mathrm{c}})$.

2.2 Waveguiding Structures on SOI

We focus on devices, more specifically photonic integrated circuits (PIC), which can be realized using the CMOS fabrication infrastructure. Fundamental building blocks are required to achieve modulation / switching, or light emission, and to realize passive routing of light and RF signals, coupling of light from and to the device.

2.2.1 Silicon Properties and Silicon-Organic Hybrid Approach

Looking at fundamental physical properties, the advantageous and disadvantageous properties of using crystalline Si as a base to build photonic waveguides (WG) and a number of (mostly) passive structures are explored in this section.

The electrical properties of silicon arise from its nature as a group IV semiconducting material. The conductivity of crystalline silicon can be controlled by simply implanting ions of groups III or V. The introduction of these dopants into the crystalline silicon creates free carriers, but also little defects in the crystal structure, with consequences for the optical properties.

Another, notably dynamic way to manipulate the conductivity of silicon is to influence the carrier density with electric fields, for instance applied using an external voltage source. A well know example is the field effect transistor (FET), often implemented as metal-oxide-semiconductor field effect transistor (MOSFET). Applying a voltage (with correct polarity) between the so-called gate electrode and the body (often connected to the source electrode) creates an electric field. This causes an accumulation of carriers between source and drain electrodes. This leads to the formation of a channel, such that a current can flow between source and drain. Note that due to the insulation of the gate with a dielectric (often an oxide), this current is controlled by a voltage and does not necessitate a current (beyond charging the gate-body capacitance) in contrast to bipolar junction transistors.

Transparency in the range of infrared (IR) wavelengths is a great advantage of silicon. Absorption is low in the most commonly used spectral transmission windows of telecommunication. These so called bands start at a wavelength of 1260 nm (O band, chosen for zero dispersion in standard single mode fiber). There are furthermore the bands E, S, C (chosen for low absorption and because erbium-doped fiber amplifiers (EDFA) are available for amplification), L, U (ending at 1675 nm). The high refractive index of 3.5 in the C band provides for high confinement of light to silicon waveguides (WG) when surrounded by standard materials (such as air, silicon oxide SiO_2). A high confinement means the light follows more closely the shape of the waveguide, which makes tighter bends possible. Hence the material is attractive to make PICs, because very dense integration is feasible. This is of advantage for making devices with low footprint, but also because a high intensity of light can be easily realized, as needed for applications in nonlinear optics.

Crystalline silicon has a bandgap of 1.1 eV, which explains the high absorption observed for wavelengths below 1.15 µm. For light propagating a distance l in silicon with a measured attenuation coefficient α_{meas} the fraction of its intensity after propagation $I_{\text{light}}(l)$ and the intensity before $I_{0,\text{ light}}$ is

$$I_{\text{light}}(l) = I_{0,\text{ light}} \exp\left(-\alpha_{\text{meas}} l\right). \tag{2.26}$$

Changing the number of free carriers by injection or depletion causes changes in the refractive index n and absorption at the same time, which is called the free carrier plasma dispersion effect. To quantify the change in transparency of silicon $\Delta\alpha_{\text{Soref}}$ one can therefore refer to a model proposed by Soref [30], [31] based on the classical Drude model. It describes the power attenuation coefficient in silicon dependent on the change of concentration of electron density ΔN_{e} and hole density ΔN_{h} (ranging between $10^{17}\ldots10^{20}\,\text{cm}^{-3}$) due to implants using the constants $C_{\text{e}} = 8.5\times10^{-22}\,\text{cm}^2$ and $C_{\text{h}} = 6.0\times10^{-22}\,\text{cm}^2$.

$$\Delta\alpha_{\text{Soref}} = C_{\text{e}}\Delta N_{\text{e}} + C_{\text{h}}\Delta N_{\text{h}} \tag{2.27}$$

The refractive index change Δn_{Soref} is related to the change in absorption by the Kramers-Kronig relation [28], and contains an empirical correction,

$$\Delta n_{\text{Soref}} = -8.8\times10^{-22}\,\text{cm}^3\Delta N_{\text{e}} + 8.5\times10^{-18}(\text{cm}^3\Delta N_{\text{h}})^{0.8}. \tag{2.28}$$

Using an empirical model from Vardanyan et al. [32] promises better accuracy of the power attenuation coefficient, for n-type (valid for $\lambda = 1.15\ldots6\,\mu\text{m}$)

$$\frac{\alpha_{\text{n, Vard}}}{\text{cm}^{-1}} = \frac{N_{\text{e}}}{\text{cm}^{-3}}10^{-18}\left(1.207\left(\frac{\lambda}{\mu\text{m}}\right)^3 - 11.70\left(\frac{\lambda}{\mu\text{m}}\right)^2 + 48.39\left(\frac{\lambda}{\mu\text{m}}\right) - 34.81\right), \tag{2.29}$$

and p-type silicon (valid for $\lambda = 1.15\ldots8\,\mu\text{m}$)

$$\frac{\alpha_{\text{p, Vard}}}{\text{cm}^{-1}} = \frac{N_{\text{h}}}{\text{cm}^{-3}}10^{-19}\left(-2.485\left(\frac{\lambda}{\mu\text{m}}\right)^3 + 65.54\left(\frac{\lambda}{\mu\text{m}}\right)^2 - 120.11\left(\frac{\lambda}{\mu\text{m}}\right) + 122.2\right). \tag{2.30}$$

However, both Soref and Vardanyan do not differentiate between free carriers created from impurity ionization and from injected free carriers. As Alloatti [33] pointed out, creating a charge accumulation layer can be the preferred approach when an increase in conductivity is desired, while an increase in optical losses has to be shunned. Injected charges cause less optical loss than free carriers from implanted ions, which represent a defect in the crystal structure. An application of this idea to optoelectronic devices, modulators in particular, has led to the patent applications [P1, P2].

The indirect nature of the bandgap of Si prevents the construction of lasers using bulk c-Si. The nonlinear optical properties of c-Si are determined by its crystal structure. As for any centro-symmetric bulk crystal, the nonlinear susceptibility contribution of $\chi^{(2)}$ and higher even orders is negligible. However, crystalline silicon has a significant real and imaginary $\chi^{(3)}$ susceptibility.

A number of applications require properties not available in silicon to enable light emission or $\chi^{(2)}$ related effects. To circumvent the limits of silicon, other materials can be combined with the PIC. Especially the class of organic materials offers a practically unlimited choice of substances and compounds with unique properties.

Silicon-organic hybrid (SOH) devices, which combine silicon and organic materials, can be constructed [8] by integrating selected organic materials with silicon. Examples for the functionality achieved with organic materials on silicon devices include:

- A strong $\chi^{(2)}$-susceptibility, discussed in section 2.4.2, to make a silicon high speed modulator.
- A strong $\chi^{(3)}$ susceptibility for optical de-multiplexing [34].
- The capability for stimulated emission, discussed in Section 2.3, to make a silicon laser.

2.2.2 SOI Waveguides

To control light and determine its propagation in PICs, the geometry of silicon and its surrounding materials is discussed here. It is of great practical value to find WG structures, which have similar Eigen-solutions (modes) of Maxwell's equations (for optical frequencies) at their input and output ports. By assuring that light is guided in the ground mode at those ports, building blocks can be defined, enabling a modular construction of the PIC. If the modes at the ports between some components do not match, mode converters have to be used.

A standard structure is an optical strip WG, see Fig. 2.1(a), which is:

- translation invariant in propagation direction (z) of the light,
- made of a rectangular cross section of c-Si ($n = 3.5$ at a wavelength of 1.55 µm, dimensions in the order of the intended wavelength, often at a width of 500 nm and height of 220 nm),
- surrounded by a material with lower refractive index, like SiO_2 ($n = 1.5$ at a wavelength of 1.55 µm) or with polymer cladding ($n = 1.7$ in Fig. 2.1(a)).

To find the modes for this high index contrast WG ($\Delta n \simeq 2$), we use CST Microwave Studio or COMSOL to numerically solve Maxwell's equations and obtain the electric field distribution, see Fig. 2.1(b). Pure transverse electric TE and transverse magnetic TM modes are not to be found in strongly guiding strip waveguides. Instead hybrid modes are called (quasi-) TE modes (have a dominant electric field E_x component) and (quasi-) TM modes (have a dominant electric field E_y component).

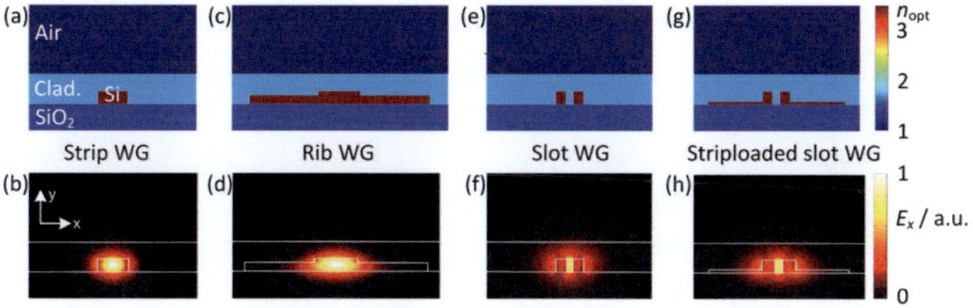

Fig. 2.1: Cross sections of optical waveguides (WG). The upper row shows the refractive index. The maximum height of the silicon part is 220 nm. The lower row shows the distribution of the electric field (the dominant E_x-component) for the quasi TE-mode. (a,b) Strip WG, (c,d) rib WG with 150 nm high Si strips on both sides of the rib, which has a width of 700 nm, (e,f) slot WG with 120 nm slot width, and (g,h) striploaded slot WG with 50 nm high silicon strips and 120 nm slot width.

To maintain compatibility with CMOS-based processes, so-called silicon-on-insulator (SOI) c-Si wafers are used here. These are manufactured to have a 750 µm thick crystalline silicon substrate, with a 2 µm thick oxide layer on which a 220 nm thick crystalline silicon layer can be structured to result in silicon waveguides. The ratios between WG height, width and SiO_2 layer thickness is chosen to avoid leaking of light into the substrate, when guided in a WG. This layer stack is by no means the only option available to make silicon photonic WGs, but a very common one, which allows for relatively small WGs (order of wavelength) and hence dense integration.

Waveguide geometries can be adapted for different purposes. Depending on the intended application, different WGs can be realized. For special purposes, i.e. dispersion engineering, periodic structures are made. Some reach a periodicity with a unit cell length below the wavelength intended to decrease the group velocity over several orders of magnitude to enhance the interaction of the light with the structure, discussed as photonic crystals for instance by [35].

Here, we discuss z-translation invariant (in propagation direction) waveguide cross sections. To enable a purpose-oriented description of the modes of these structures, a number of useful definitions are given:

The effective refractive index n_{eff} for a waveguide or its propagation constant β need to be determined numerically for the waveguide profiles presented below and depends on the angular frequency ω of the light.

The interaction of the light in a WG with a given region (modal overlap integral with a given region) of the waveguide can be quantified with the interaction factor [36]. The same coordinate system as introduced in Fig. 2.1 is used. The refractive index of the region of interest (integration domain of the numerator integral), is described by n_{reg}. Z_0 is the vacuum impedance, $\mathbf{e}_z$ the unit vector in z-direction. The symbol * denotes the complex conjugate. Here, the interaction factor Γ, also sometimes called confinement factor, is defined as

$$\Gamma = \frac{\int\limits_{\text{reg}} \frac{n_{\text{reg}}}{Z_0} \left| \tilde{\mathbf{E}}_{\text{t}}\left(x, y, \omega_{\text{c}}\right) \right|^2 \mathrm{d}x\mathrm{d}y}{\int\limits_{\text{all}} \Re\left\{ \tilde{\mathbf{E}}_{\text{t}}\left(x, y, \omega_{\text{c}}\right) \times \tilde{\mathbf{H}}_{\text{t}}^{*}\left(x, y, \omega_{\text{c}}\right) \right\} \cdot \mathbf{e}_z \mathrm{d}x\mathrm{d}y}. \tag{2.31}$$

The integration domain of the denominator integral is the entire waveguide cross section. A large interaction factor for a certain WG region means that a respectively large amount of light is guided in this WG region. The interaction factor allows for a convenient linear approximation of the effect of changing the refractive index Δn_{reg} of a material/part of a WG on the resulting change of effective refractive index Δn_{eff} of the WG.

$$\Delta n_{\text{eff}} \simeq \Gamma \Delta n_{\text{reg}} \tag{2.32}$$

Dependent on the application it can be useful to integrate only over the effective components of the electric field $\tilde{\mathbf{E}}_{\text{t}}\left(x, y, \omega_{\text{c}}\right)$ in the numerator. For instance, if only the x-component is relevant for an interaction, then $\tilde{\mathbf{E}}_{\text{t}}\left(x, y, \omega_{\text{c}}\right) \cdot \mathbf{e}_x$ is integrated. In this case, the field interaction factor Γ_{Ex} for the x-component of the electric field is

$$\Gamma_{Ex} = \frac{\int\limits_{\text{reg}} \frac{n_{\text{reg}}}{Z_0} \left| \tilde{\mathbf{E}}_{\text{t}}(x, y, \omega_{\text{c}}) \cdot \mathbf{e}_x \right|^2 \text{d}x\text{d}y}{\int\limits_{\text{all}} \Re\left\{ \tilde{\mathbf{E}}_{\text{t}}(x, y, \omega_{\text{c}}) \times \tilde{\mathbf{H}}_{\text{t}}^*(x, y, \omega_{\text{c}}) \right\} \cdot \mathbf{e}_z \text{d}x\text{d}y}. \tag{2.33}$$

The change of refractive index can be calculated similar to Eq. (2.32).

$$\Delta n_{\text{eff}} \simeq \Gamma_{Ex} \Delta n_{\text{reg}} \tag{2.34}$$

For effects depending linearly on intensity, a description of the light distribution by the effective mode cross section A_{eff} is useful, which is defined here [29] as

$$A_{\text{eff}} = \frac{\left(\int\limits_{\text{all}} \Re\left\{ \tilde{\mathbf{E}}_{\text{t}}(x, y, \omega_{\text{c}}) \times \tilde{\mathbf{H}}_{\text{t}}^*(x, y, \omega_{\text{c}}) \right\} \cdot \mathbf{e}_z \text{d}x\text{d}y \right)^2}{\int\limits_{\text{reg}} \frac{n_{\text{reg}}^2}{Z_0^2} \left| \tilde{\mathbf{E}}_{\text{t}}(x, y, \omega_{\text{c}}) \right|^4 \text{d}x\text{d}y}. \tag{2.35}$$

A smaller effective mode cross section means that the amount of light guided in this region is more concentrated, i.e., the intensity is higher, which can be of advantage in a number of nonlinear applications.

Another measure proposed to describe the spatial extent in x-direction of the TE-mode for a region of a WG is the mode field diameter (MFD). Similar to a common method to derive the MFD of a beam, we calculate it as a second moment width along x-direction, for a definition of moments see Appendix A.2. The difference to the common definition is that we only consider part of the cross section, i.e. a particular region. The purpose is to derive an effective width for the part of light which is guided in that region. To calculate it, we need the first moment $\overline{x}$, which will be zero as we only consider WGs which are symmetric in the y-z-plane.

$$\overline{x} = \frac{\int\limits_{\text{reg}} \frac{n_{\text{reg}}}{Z_0} \left| \tilde{\mathbf{E}}_{\text{t}}(x, y, \omega_{\text{c}}) \right|^2 \text{x } \text{d}x\text{d}y}{\int\limits_{\text{all}} \Re\left\{ \tilde{\mathbf{E}}_{\text{t}}(x, y, \omega_{\text{c}}) \times \tilde{\mathbf{H}}_{\text{t}}^*(x, y, \omega_{\text{c}}) \right\} \cdot \mathbf{e}_z \text{d}x\text{d}y} \tag{2.36}$$

The effective mode field diameter is defined here as

$$\mathrm{MFD}_x = 4\sqrt{\frac{\int\limits_{\mathrm{reg}} \frac{n_{\mathrm{reg}}}{Z_0} \left|\tilde{\mathbf{E}}_{\mathrm{t}}(x,y,\omega_{\mathrm{c}})\right|^2 (x-\overline{x})^2 \,\mathrm{d}x\mathrm{d}y}{\int\limits_{\mathrm{all}} \Re\left\{\tilde{\mathbf{E}}_{\mathrm{t}}(x,y,\omega_{\mathrm{c}}) \times \tilde{\mathbf{H}}_{\mathrm{t}}^{*}(x,y,\omega_{\mathrm{c}})\right\} \cdot \mathbf{e}_z \mathrm{d}x\mathrm{d}y}}. \tag{2.37}$$

A large MFD means a larger extension of the mode in the considered region along x-direction.

While the maximum height of the c-Si part of WGs is fixed to 220 nm by the employed wafers in this work, a number of geometrically different WG cross sections can be considered: (a) to transport light efficiently over long distances on the chip and (b) to allow efficient interaction of the light with the material, while maintaining compatibility to CMOS fabrication processes at the same time.

Efficient on-chip long distance travel of light

The strip waveguide presented above presents the standard solution to guide light between different functional components on a PIC. Its fundamental mode can be converted to the fundamental modes of other waveguide cross sections, e.g. by adiabatic tapers. A typical width of 450…500 nm is chosen to make tight bends (radius of 5 µm, [37]) without needing to change the WG's width. An example of its electric field distribution (dominant x-component) is presented in Fig. 2.1(b). However, in this figure it becomes apparent that the intensity of light at the sidewalls of the Si strip is relatively high, which will result in scattering and thus propagation loss of typically 2…3 dB/cm (for WGs from DUV). Recently, the rib waveguide shown in Fig. 2.1(c) has been proposed and demonstrated [38] to exhibit lowest loss of 0.7 dB/cm. Its large rib width of 700 nm, and even larger Si slab (the Si strip on both sides with a height of 150 nm) width of 3 µm confines the light to the largest part in the Si WG itself. Crucially, the area of the sidewalls next to the rib is strongly reduced, see Fig. 2.1(d), thus scattering minimized.

Efficient on-chip light-material interaction

In addition to the strip WG presented above, waveguides which maximize the interaction of light with the cladding are of interest for a number of applications, especially if the properties of an organic cladding are to be exploited using the SOH approach. The confinement of light to a strip wave-waveguide can be reduced by decreasing its width, which reduces the effective refractive index and thus increases the overlap with the cladding. A lower limit of this solution is imposed by the substrate in close proximity, which will cause leaking of light from WGs with too small widths. The slot waveguide depicted in Fig. 2.1(e) increases the interaction of light with the cladding. It consists of two narrow strips, also called Si rails, which are placed parallel and close to each other. The components of the electric displacement field $\mathcal{D}(\mathbf{r},t)$ normal to the interface are continuous across the interface of cladding and silicon,

$$\mathbf{n}\cdot\mathcal{D}_{\mathrm{Si}}(\mathbf{r},t)=\mathbf{n}\cdot\mathcal{D}_{\mathrm{Clad}}(\mathbf{r},t). \tag{2.38}$$

This leads to a strong enhancement of the E_x-component of the electric field in the slot, see Fig. 2.1(f), and thus to strong interaction of the guided light with the cladding material. By adding Si strips of low height (e.g. 45…70 nm) to the rails, as depicted in Fig. 2.1(g), a strip-loaded slot waveguide (also called socket waveguide) is created. The confinement of light in the cladding remains high, Fig. 2.1(h). By doping the strips, an electrical contact can be established from the rails to the other end of the Si strip. At this end metal electrodes can be placed, without risking an attenuation of the guided light by the metal.

A standard concern is to achieve low-loss propagation. Losses are caused in part by absorption in the bulk crystal, Eq. (2.27), caused in part by surface states [39], but primarily caused by scattering on non-ideally smooth material interfaces [36]. While the surface of an SOI wafer comes close to be atomically flat, the roughness of WG sidewalls is mostly determined by the fabrication methods. Especially the lithographic definition of waveguides is a major issue, because the roughness of the chromium mask is eventually transferred to the silicon WG's sidewalls by a dry etching. The amount of scattering strongly depends on the dimensions of surface roughness [36] and also on the refractive index contrast at the material interface. If a lithographic resist pattern is well defined, smoother waveguide sidewalls with less scattering losses will result. This is the case when using e-beam lithography compared to deep UV lithography (at a wavelength of 193 nm). The latter nevertheless presents the only viable option for mass production.

For silicon WGs with air cladding, such as of interest for bio-sensing applications, it has been suggested and demonstrated in a collaboration with

Honkanen et al. [J13], [J12] that an additional material layer between Si and air can reduce losses. A deposition technique called atomic layer deposition (ALD) is employed. It creates precise, smooth, ultra-thin layers (several tens of nm thick coating) of TiO_2. As shown in the next figure, variations on the silicon surface will not be completely transferred to new layers grown with TiO_2, but will be gradually lost with increasing thickness. The refractive index of TiO_2 is $n = 2.27$ at a wavelength of 1.55 μm. That means the original material interface Si-air is substituted with two interfaces, Si-TiO_2 (same roughness as Si-air, but reduced index contrast) and TiO_2-air (strongly reduced roughness compared to Si-air, slightly weaker index contrast). As a result scattering losses are reduced.

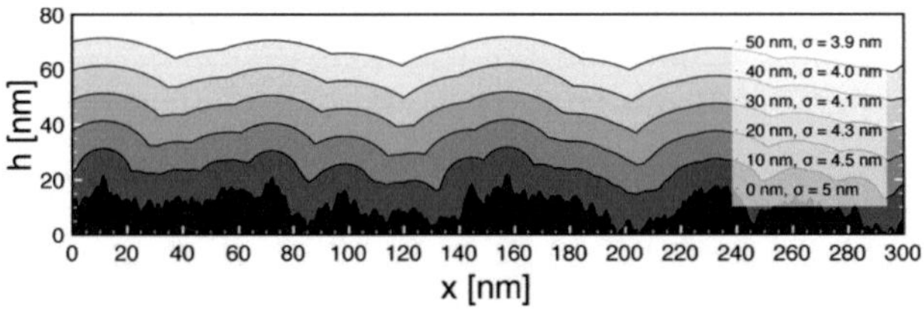

Fig. 2.2: Simulation TiO2 grown by atomic layer deposition (ALD) on a rough Si surface. Darkest grey depicts the Si with a surface roughness which is exponentially distributed (correlation length is 15 nm) and has an effective (RMS) roughness of $\sigma_{RMS} = 5$ nm. The deposition of 10 nm thick layers of TiO2 shows that the resulting effective roughness is reduced of each layer. Image source: Fig. 1 in [J13], ©OSA 2011.

Anisotropic wet etching presents another method to correct for imperfections of the lithographic mask and is discussed by Palmer et al. [C25]. This method relies on the different etch velocities for different Si crystal planes, as the name suggests. On the one hand, exposing crystal planes provides atomically flat surfaces (up to a limited spatial extend), i.e. reduces scattering to a minimum. On the other hand, any structure which is to be produced with this method has to have its WG sidewalls aligned according to the crystal planes. This significantly restricts the possible shapes to form photonic structures and requires further effort to develop building blocks compatible to anisotropic wet etching.

2.2.3 SOI Standard Structures

Numerous combinations of the waveguide types presented in the preceding section can form a PIC. More functionality is provided by standard structures discussed here.

Coupling light to and from the chip can be done by utilizing inverted[10] tapers to increase the mode field diameter of the light guided from the Si chip to a small-core fiber (mode field diameter e.g. 3 μm). Another approach suited for prototyping and used throughout this work is based on grating couplers, i.e. Bragg gratings defined by shallow trenches on top of a wide strip waveguide to diffract light from the WG into a standard single mode fiber (SSMF, mode field diameter 10 μm). Typical losses are 3…5 dB/coupler [40]. The use of p-Si over-claddings can significantly reduce the losses further.

Transitions from strip to slot and strip-loaded slot waveguides have been designed, fabricated and characterized by Palmer et al. [J8]. In essence: an adiabatic transformation from the quasi TE mode of the strip is made to the TE mode of the slot waveguide, in such a way that both rails remain electrically insulated.

Mach-Zehnder Interferometers (MZI) can be used to make spectral filters or a conversion of electrical signals into optical signals. The MZI is constructed in this work by combing waveguides with multi-mode interference couplers (MMI) (directional couplers present another option to make splitters and combiners, as demonstrated elsewhere [41]). Here, we consider an MZI with a delay $\Delta\tau$ in one arm and two phase shifters φ_1 and φ_2, one in each arm, as shown in the next figure. All optical parts are depicted in blue. It is assumed that the polarization of the light remains unaffected by the device, i.e., no polarization dependent effects are considered here, because the MZMs are operated with quasi-TE modes all the time. The phase shifters are operated by applying an effective voltage $V_{\mathrm{S,eff}}$ between the ground-signal-ground (GSG) electrodes. All electrical parts are depicted in orange or red. The particular implementation of the phase shifter and its response to a voltage is defined later.

[10] Inverted tapers are made from strip WGs, where the width (and ideally also height) of the Si strip is tapered down towards the end of the WG, such that the light coming from the PIC is less and less confined to the Si strip and the mode field diameter is growing. Light can propagate on the same structure in the other direction as well.

Here we only assume the resulting phase shifts as linear functions $\varphi_1\left(V_{\mathrm{S,eff}}\right)$ and $\varphi_2\left(V_{\mathrm{S,eff}}\right)$.

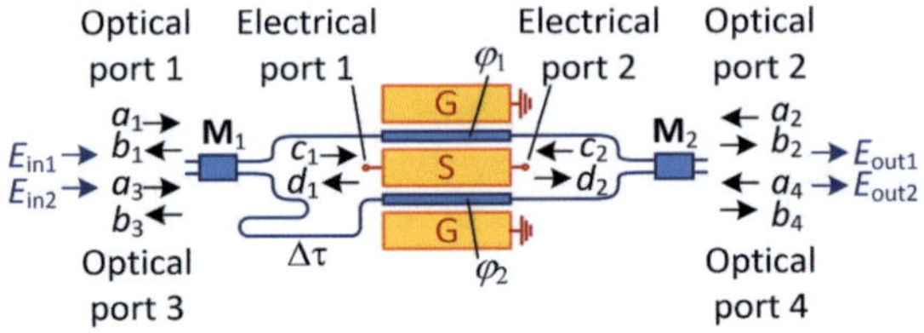

Fig. 2.3: Delay interferometer (DI) composed of 2 multi-mode interference couplers (MMI), a delay line introducing $\Delta\tau$ and a phase shifters in both arms. The signals $a_1,\ldots,a_4$ are sent into the respective ports, while the signals $b_1,\ldots,b_4$ will then come out at the respective ports. During operation electric fields $E_{\mathrm{in1,in2}}$ are sent into the left MMI and the fields $E_{\mathrm{out1,out2}}$ result.

Optical-to-optical scattering matrix $\mathbf{S}_{\mathrm{OO}}$

We first consider the optical behavior of this device in a four port model. The reaction of the device on optical input signals $\mathbf{a}=(a_1,\ldots,a_4)^{\mathrm{T}}$ (normalized wave amplitudes, such that $|\mathbf{a}|^2$ represents the incoming optical power) at each port can be described using a four-port optical-to-optical scattering matrix $\mathbf{S}_{\mathrm{OO}}$. One can describe the optical output signals $\mathbf{b}=(b_1,\ldots,b_4)^{\mathrm{T}}$ (normalized wave amplitudes, such that $|\mathbf{b}|^2$ represents the outgoing optical power) at each port using

$$\mathbf{b}=\begin{pmatrix} b_1 \\ b_2 \\ b_3 \\ b_4 \end{pmatrix},\quad \mathbf{S}_{\mathrm{OO}}=\begin{pmatrix} S_{11,\mathrm{OO}} & \cdots & S_{14,\mathrm{OO}} \\ \vdots & \ddots & \vdots \\ S_{41,\mathrm{OO}} & \cdots & S_{44,\mathrm{OO}} \end{pmatrix},\quad \mathbf{a}=\begin{pmatrix} a_1 \\ a_2 \\ a_3 \\ a_4 \end{pmatrix} \tag{2.39}$$

to be

$$\mathbf{b}=\mathbf{S}_{\mathrm{OO}}\mathbf{a}\,. \tag{2.40}$$

For example, light could enter as a_1 at the first port and if there is no other signal present, one could easily predict the outputs at each port $b_1,\ldots,b_4$ from the first column of the scattering matrix. Here, we decide to use only the ports 1 and 2 for input. We further assume a loss-less and reflection-less device.

Hence, we only need the elements $S_{21,\mathrm{OO}}$, $S_{41,\mathrm{OO}}$, $S_{23,\mathrm{OO}}$, $S_{43,\mathrm{OO}}$ to form a new transfer matrix $\tilde{\mathbf{S}}_{\mathrm{OO,MZI}}$,

$$\tilde{\mathbf{S}}_{\mathrm{OO,MZI}} = \begin{pmatrix} S_{21,\mathrm{OO}} & S_{23,\mathrm{OO}} \\ S_{41,\mathrm{OO}} & S_{43,\mathrm{OO}} \end{pmatrix}. \tag{2.41}$$

This transfer matrix can be used to calculate the electric fields $\tilde{E}_{\mathrm{out1}}$, $\tilde{E}_{\mathrm{out2}}$ in the frequency domain at the output from the given input $\tilde{E}_{\mathrm{in1}}$, $\tilde{E}_{\mathrm{in2}}$, also in the frequency domain.

$$\begin{pmatrix} \tilde{E}_{\mathrm{out1}}(\omega) \\ \tilde{E}_{\mathrm{out2}}(\omega) \end{pmatrix} = \tilde{\mathbf{S}}_{\mathrm{MZI,OO}} \begin{pmatrix} \tilde{E}_{\mathrm{in1}}(\omega) \\ \tilde{E}_{\mathrm{in2}}(\omega) \end{pmatrix} \tag{2.42}$$

The time delay calls for a description in the time domain ($*$ denotes a component-wise convolution)

$$\begin{pmatrix} E_{\mathrm{out1}}(t) \\ E_{\mathrm{out2}}(t) \end{pmatrix} = \mathbf{S}_{\mathrm{MZI,\,OO}} * \begin{pmatrix} E_{\mathrm{in1}}(t) \\ E_{\mathrm{in2}}(t) \end{pmatrix} \tag{2.43}$$

In this formalism, the transfer matrix of the differently delayed arms is

$$\mathbf{D} = \begin{pmatrix} \delta(t) & 0 \\ 0 & \delta(t-\Delta\tau) \end{pmatrix}. \tag{2.44}$$

The transfer matrix of the phase shifter[11] in each arm is

$$\mathbf{\Phi} = \begin{pmatrix} \exp(\mathrm{j}\varphi_1) & 0 \\ 0 & \exp(\mathrm{j}\varphi_2) \end{pmatrix}. \tag{2.45}$$

The transfer matrices of the 2-by-2 MMIs (3 dB splitters/combiner) can be written as

$$\mathbf{M}_1 = \frac{1}{\sqrt{2}}\begin{pmatrix} 1 & \mathrm{j} \\ \mathrm{j} & 1 \end{pmatrix},\ \mathbf{M}_2 = \frac{1}{\sqrt{2}}\begin{pmatrix} 1 & \mathrm{j} \\ \mathrm{j} & 1 \end{pmatrix}. \tag{2.46}$$

---

[11] For example, applying a phase shift to an electric field as in Eq. (2.14) results in

$$E_{\mathrm{out}}(z,t) = E_{\mathrm{in}}(z,t)\exp(\mathrm{j}\varphi_1)$$
$$\propto \exp\left(\mathrm{j}(\omega_{\mathrm{c}}t-\beta z)\right)\exp(\mathrm{j}\varphi_1) = \exp\left(\mathrm{j}(\omega_{\mathrm{c}}t-\beta z+\varphi_1)\right).$$

The product of all transfer matrices gives the total matrix for the MZI:

$$\begin{pmatrix} E_{\text{out1}}(t) \\ E_{\text{out2}}(t) \end{pmatrix} = \mathbf{S}_{\text{MZI,OO}} * \begin{pmatrix} E_{\text{in1}}(t) \\ E_{\text{in2}}(t) \end{pmatrix} = \mathbf{M}_2 \mathbf{\Phi} \mathbf{D} \mathbf{M}_1 * \begin{pmatrix} E_{\text{in1}}(t) \\ E_{\text{in2}}(t) \end{pmatrix} \tag{2.47}$$

$$\mathbf{S}_{\text{MZI,OO}} =$$

$$\frac{1}{2}\begin{pmatrix} \exp(\mathrm{j}\varphi_1)\delta(t) - \exp(\mathrm{j}\varphi_2)\delta(t-\Delta\tau) & \mathrm{j}\exp(\mathrm{j}\varphi_1)\delta(t) + \mathrm{j}\exp(\mathrm{j}\varphi_2)\delta(t-\Delta\tau) \\ \mathrm{j}\exp(\mathrm{j}\varphi_1)\delta(t) + \mathrm{j}\exp(\mathrm{j}\varphi_2)\delta(t-\Delta\tau) & -\exp(\mathrm{j}\varphi_1)\delta(t) + \exp(\mathrm{j}\varphi_2)\delta(t-\Delta\tau) \end{pmatrix} \tag{2.48}$$

Using the Fourier transform this convolution can be written as a product in frequency space, as in Eq. (2.42) with

$$\tilde{\mathbf{S}}_{\text{MZI,OO}} = \frac{1}{2}\begin{pmatrix} \exp(\mathrm{j}\varphi_1) - \exp(\mathrm{j}\varphi_2\text{-}\mathrm{j}\omega\Delta\tau) & \mathrm{j}\exp(\mathrm{j}\varphi_1) + \mathrm{j}\exp(\mathrm{j}\varphi_2\text{-}\mathrm{j}\omega\Delta\tau) \\ \mathrm{j}\exp(\mathrm{j}\varphi_1) + \mathrm{j}\exp(\mathrm{j}\varphi_2\text{-}\mathrm{j}\omega\Delta\tau) & -\exp(\mathrm{j}\varphi_1) + \exp(\mathrm{j}\varphi_2\text{-}\mathrm{j}\omega\Delta\tau) \end{pmatrix}. \tag{2.49}$$

Some special cases can be considered. From the previous equation the frequency responses for the device $\tilde{h}_1(\omega)$, $\tilde{h}_2(\omega)$ in case of the input at port 1 can be extracted.

$$\begin{aligned} \tilde{E}_{\text{out1}}(\omega) &= \tilde{h}_{\text{out1,in1}}(\omega)\tilde{E}_{\text{in1}}(\omega) = \tilde{h}_1(\omega)\tilde{E}_{\text{in1}}(\omega) \\ \tilde{E}_{\text{out2}}(\omega) &= \tilde{h}_{\text{out2,in1}}(\omega)\tilde{E}_{\text{in1}}(\omega) = \tilde{h}_2(\omega)\tilde{E}_{\text{in1}}(\omega) \end{aligned} \tag{2.50}$$

Using an identity in Eq. (7.3) from the Appendix $\tilde{h}_1(\omega)$ and the respective scattering matrix element of $\mathbf{S}_{\text{OO}}$ can be written as

$$S_{21,\text{OO}}(\omega) = \tilde{h}_1(\omega) = \underbrace{-\exp\left(\frac{\mathrm{j}(\varphi_1+\varphi_2-\omega\Delta\tau)}{2}\right)}_{\text{Phase modulation}} \underbrace{\sin\left(\frac{(\varphi_1-\varphi_2+\omega\Delta\tau)}{2}\right)}_{\text{Amplitude modulation}}. \tag{2.51}$$

Note the dependence on frequency. Correspondingly, $\tilde{h}_2(\omega)$ is

$$S_{43,\text{OO}}(\omega) = \tilde{h}_2(\omega) = \underbrace{\exp\left(\frac{\mathrm{j}(\varphi_1+\varphi_2-\omega\Delta\tau)}{2}\right)}_{\text{Phase modulation}} \underbrace{\cos\left(\frac{(\varphi_1-\varphi_2+\omega\Delta\tau)}{2}\right)}_{\text{Amplitude modulation}}. \tag{2.52}$$

Hence, the absolute square of the frequency responses for both output ports are

$$\left|\tilde{h}_1(\omega)\right|^2 = \sin^2\left(\frac{\omega\Delta\tau+\varphi_1-\varphi_2}{2}\right) \text{ and } \left|\tilde{h}_2(\omega)\right|^2 = \cos^2\left(\frac{\omega\Delta\tau+\varphi_1-\varphi_2}{2}\right). \tag{2.53}$$

The time delay of the light propagating in the elongated arm can be easily derived using the group velocity $v_{\mathrm{g}} = c / n_{\mathrm{g}}$ or the group refractive index n_{g}, which in turn is supplied by numerical simulations of the rib waveguide cross section around the center wavelength λ_{c}.

$$n_{\mathrm{g}}(\lambda_{\mathrm{c}}) = n(\lambda_{\mathrm{c}}) - \lambda_{\mathrm{c}} \left. \frac{\mathrm{d}n(\lambda)}{\mathrm{d}\lambda} \right|_{\lambda=\lambda_{\mathrm{c}}} \tag{2.54}$$

Electrical-to-electrical scattering matrix $\mathbf{S}_{\mathrm{EE}}$

Now we consider the electrical behavior of this device in a two port model. The reaction of the device on electrical input signals $\mathbf{c}=(c_1, c_2)^{\mathrm{T}}$ (normalized wave amplitudes, such that $|\mathbf{c}|^2$ represents the incoming electrical power) at each port can be described using a two-port electrical-to-electrical scattering matrix $\mathbf{S}_{\mathrm{OO}}$. One can describe the electrical output signals $\mathbf{d}=(d_1,d_2)^{\mathrm{T}}$ (normalized wave amplitudes, such that $|\mathbf{d}|^2$ represents the outgoing electrical power) at each port using

$$\mathbf{d} = \begin{pmatrix} d_1 \\ d_2 \end{pmatrix}, \ \mathbf{S}_{\mathrm{EE}} = \begin{pmatrix} S_{11,\mathrm{EE}} & S_{12,\mathrm{EE}} \\ S_{21,\mathrm{EE}} & S_{22,\mathrm{EE}} \end{pmatrix}, \ \mathbf{c} = \begin{pmatrix} c_1 \\ c_2 \end{pmatrix} \tag{2.55}$$

to be

$$\mathbf{d} = \mathbf{S}_{\mathrm{EE}} \mathbf{c} \tag{2.56}$$

The scattering matrix elements are dependent on the RF frequency ω_{RF}. The element $S_{21,\mathrm{EE}}(\omega_{\mathrm{RF}})$ represents the transmission which can be measured at the electrical port 2 after sending in a signal at the electrical port 1. The element $S_{11,\mathrm{EE}}(\omega_{\mathrm{RF}})$ represents reflections of incoming signals at port 1.

Electrical-to-optical scattering matrix $\mathbf{S}_{\mathrm{EO}}$

Assuming a_1 is a non-zero constant and a_1, a_1, a_1 are all zero, the reaction of the device on the electrical input signal c_1 (normalized amplitude, such that $|c_1|^2$ represents the incoming electrical power) at the electrical port 1 can be described using a three port model with an electrical-to-optical scattering matrix $\mathbf{S}_{\mathrm{EO}}$. One can derive the optical output signals $\mathbf{b}_{\mathrm{EO}} = (b_2,b_4)^{\mathrm{T}}$ (normalized amplitudes, such that $|\mathbf{b}_{\mathrm{EO}}|^2$ represents the outgoing optical power) at the optical ports 2, 4 using

$$\mathbf{b}_{\text{EO}} = \begin{pmatrix} b_2 \\ b_4 \end{pmatrix},\ \mathbf{S}_{\text{EO}} = \begin{pmatrix} S_{21,\text{EO}} \\ S_{41,\text{EO}} \end{pmatrix},\ c_1 \tag{2.57}$$

to be

$$\mathbf{b}_{\text{EO}} = c_1 \mathbf{S}_{\text{EO}}. \tag{2.58}$$

The scattering matrix element $S_{21,\text{EO}}$ allows calculating the optical response of an MZI at the optical port 2 for an electrical input signal at the electrical input port 1. In general, a response dependent on the RF frequency can be expected, i.e., one can write $S_{21,\text{EO}}(\omega_{\text{RF}})$.

Optical-to-electrical scattering matrix $\mathbf{S}_{\text{OE}}$

In case of a detector, a change in light intensity (corresponding to amplitude a_1) at the optical input port 1 will result in an electrical output of amplitude d_2 at the electrical port 2. This response can be quantified with the detector responsivity R_{D} using the elementary charge e, the quantum efficiency η_{QE} and the energy of the incident photons $\hbar\omega$ ($\hbar$ is the reduced Planck constant, see Appendix) as

$$R_{\text{D}} = \eta_{\text{QE}} \frac{e}{\hbar\omega} \text{ and hence } S_{21,\text{OE}}(\omega) \propto \eta_{\text{QE}} \frac{e}{\hbar\omega}. \tag{2.59}$$

Mach-Zehnder Modulators and IQ Modulators [42] can be built using MZIs with fast phase shifters. We first consider an MZI as in Fig. 2.3, but without the delay ($\Delta\tau = 0$). This is called a *Mach-Zehnder modulator (MZM)*.

If both phase shifters are driven to provide the same phase shift, i.e., $\varphi_1 = \varphi_2 = \varphi$ and $\varphi_1 - \varphi_2 = 0$ then Eqs. (2.51) and (2.52) can be simplified to

$$\tilde{h}_{1,\text{push-push}}(\omega) = 0 \text{ and } \tilde{h}_{2,\text{push-push}}(\omega) = \exp(\mathrm{j}\varphi). \tag{2.60}$$

A pure phase modulation at port 2 results (which can be equally obtained at port 4, if phase difference of π is chosen). This mode of operation is called *push-push operation.*

Alternatively, if both phase shifters are driven with signals of opposite sign, i.e., $\varphi_1 = -\varphi_2 = \varphi$ and $\varphi_1 - \varphi_2 = 0$, a pure amplitude modulation results at port 2. This mode of operation is called *push-pull operation* around the minimum transmission point, see also Fig. 2.4.

$$\tilde{h}_{1,\text{push-pull}}(\omega) = -\sin(\varphi) \text{ and } \tilde{h}_{2,\text{push-pull}}(\omega) = \cos(\varphi) \tag{2.61}$$

An additional difference of π/2 between both phase shifters results in an operating point called quadrature point, around which intensity modulation can be done.

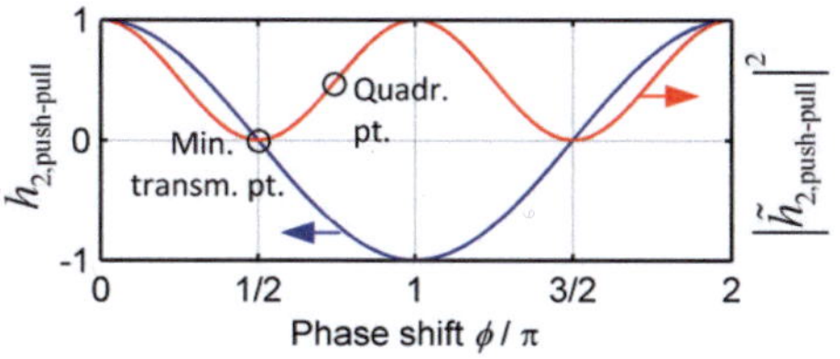

Fig. 2.4: Transfer functions for field $\tilde{h}_{2,\text{push-pull}}(\omega)$ (in blue) and intensity $|\tilde{h}_{2,\text{push-pull}}(\omega)|^2$ (in red) of MZM in push-pull operation. An operation at the minimum transmission point results in amplitude modulation, while the quadrature point is suited for intensity modulation.

IQ modulators can be built by nesting two MZMs in a large MZI, as shown in the Fig. 2.5(a). The so-called *in-phase path* I joins the splitter as it is, while the other arm accumulates an additional phase difference of π/2 (either by a slight detour, or by another phase shifter, not shown here). This second arm is called the *quadrature-phase* path Q.

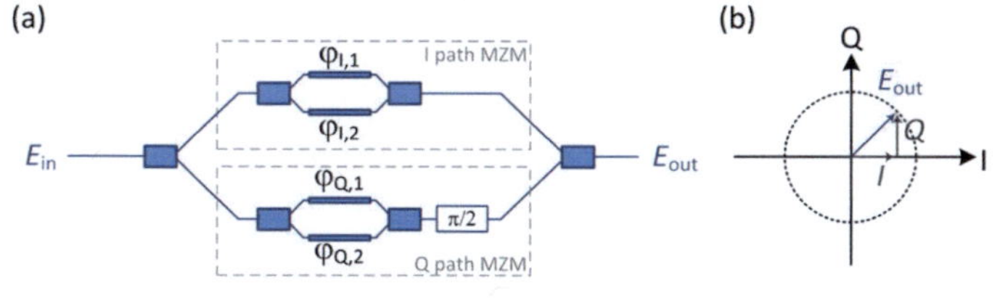

Fig. 2.5: Schematic depiction of IQ modulator and operation principle. (a) Two Mach-Zehnder modulators (MZM) are placed in a Mach-Zehnder interferometer (MZI). Both MZM are operated in push-pull mode at their minimum transmission point. Hence, each MZM for itself performs a pure amplitude modulation. An cw electric field E_{in} is split by a 3-dB splitter (1x2 MMI) at the input of the IQ modulator and each part modulated by an MZM. The upper arm joins a 3-dB combiner (1x2 MMI). The lower arm joins this combiner as well, but is phase-shifted by π/2 before. (b) Thus the resulting output field with amplitude E_{out} is the superposition of the I and Q fields, as depicted in the constellation diagram (also called IQ-diagram). Any point in this constellation diagram can be reached by choosing the right amplitudes at the MZMs.

To calculate the transfer function it is assumed that the I-path MZM and the Q-path MZM are operated in push-pull mode ($\varphi_{I,1} = -\varphi_{I,2} = \varphi_I, \varphi_{Q,1} = -\varphi_{Q,2} = \varphi_Q$)

at their minimum transmission operating point. The transfer function of each MZM we write as

$$\tilde{h}_{\mathrm{I}}(\omega)=\sin\left(\varphi_{\mathrm{I}}\right) \text{ and } \tilde{h}_{\mathrm{Q}}(\omega)=\sin\left(\varphi_{\mathrm{Q}}\right). \tag{2.62}$$

The transfer function of the IQ modulator (using 1×2 MMIs) for the electric field is

$$\begin{aligned}\tilde{h}_{\mathrm{IQ}}(\omega)&=\frac{1}{\sqrt{2}}\begin{pmatrix}1 & 1\end{pmatrix}\begin{pmatrix}\sin\left(\varphi_{\mathrm{I}}\right) & 0\\ 0 & \mathrm{j}\sin\left(\varphi_{\mathrm{Q}}\right)\end{pmatrix}\frac{1}{\sqrt{2}}\begin{pmatrix}1\\1\end{pmatrix}\\ &=\frac{1}{2}\left(\sin\left(\varphi_{\mathrm{I}}\right)+\mathrm{j}\sin\left(\varphi_{\mathrm{Q}}\right)\right).\end{aligned} \tag{2.63}$$

That means any input complex field will experience an amplitude modulation of the real and respectively imaginary part. Any point in a constellation diagram can be reached by choosing the right amplitudes at the MZMs, see Fig. 2.5(b). During detection, it will depend on a reference phase, which part of the signal is to be projected on the real axis and therefore measured. Both, I and Q part of the signal have to be measured separately.

2.3 Stimulated Emission on SOI

Making silicon lasers is a formidable challenge due to silicon's indirect bandgap. A number of hybrid integration schemes using inorganic materials is pursued elsewhere [20], [43]. In this work, we investigate the suitability of integrating a cladding of dye molecules on silicon waveguides to make a laser on silicon.

Looking at lasers realized outside the domain of integrated optics, dye lasers once were the technology of choice to reach high peak powers, but came out of fashion, due to a number of issues. This section follows closely the arguments summarized in [44] and will serve to explain the physical background of the merits and demerits of dye lasers.

2.3.1 General Dye Properties

Dye molecules usually strongly interact with light of certain wavelengths (they have a color), an interaction which is determined by their chromophores. This part of the dye molecules is often a conjugated pi-bond system, i.e. these organic molecules[12] have large, distributed orbitals (encompassing many carbon atoms). These orbitals can easily be designed to enable electronic transitions at the desired frequencies, by changing the length/structure of the conjugated pi bond system or influencing it with other molecular groups. Besides showing tunable absorption properties, a subset constituted of laser dyes is also suited for stimulated emission.

From a historical point of view, classical liquid dye lasers have been investigated at least since 1966, mainly because of their widely tunable emission wavelength range and the possibility to go to another wavelength range by switching the gain medium, which in its liquid form is easy to exchange. Even a mixture of dyes can be made, to further increase the available emission wavelength range without the need of effecting a gain medium change. The fact that dyes are contained in a liquid solution also simplified variations of the concentration to adjust absorption and emission, or replacing a damaged gain medium.

A major disadvantage is rooted in the chemical stability of the dyes. Pump light in some instances can destroy molecules by supplying activation energy to start undesired chemical reactions. This is also known as bleaching (loss of color because of chromophore destruction), due to its limited photo-chemical stability. Dye laser operation required in some cases a frequent replacement of the dye solution. In addition, continuous wave (cw) emission is hard to achieve, which can be explained with a typical, simplified energy level diagram, as depicted in the next figure [45].

The simplified energy level diagram comprises two manifolds of electronic states due to the conjugated pi-bond system. So called singlet states S_0 (ground state), S_1, in which the considered electron system has total spin of 0, and triplet states T_1, T_2, in which the total spin is 1, see p. 4 in [46]. The electronic levels are broadened by vibrational states of the molecule, with energy differences much smaller than for electronic states. In addition, there are rotational states, causing an even finer energy level splitting. The set of vibrational and rotational states is very dense and extends each electronic level into a quasi-continuous

[12] Organic molecules are molecules that contain carbon. Carbides, carbonates, carbon oxides, cyanides and carbon allotropes (diamond, graphene, etc.) are considered as inorganic compounds.

band. When looking at an ensemble of dye molecules, even in thermal equilib-equilibrium not all electrons are in the vibrational, rotational ground state. The probability to find an electron in a certain state by can be modeled with the Boltzmann distribution.

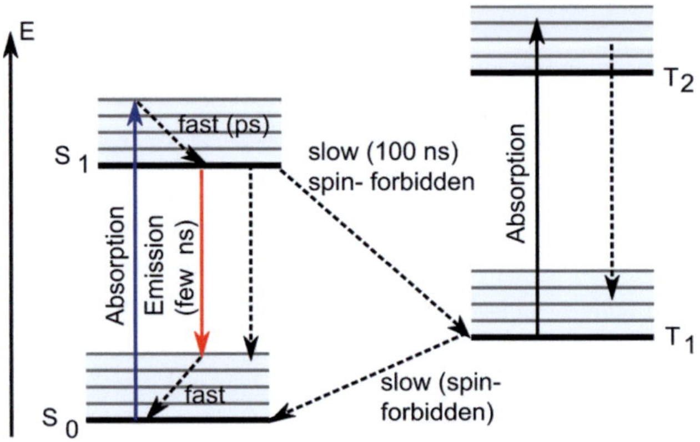

Fig. 2.6: Typical, simplified energy level diagram of a dye molecule. Solid lines denote transitions involving photons. In radiative transitions a photon is emitted. Perforated lines denote non-radiative transitions. Absorption of pump light (blue), fast relaxation within vibrational and rotational states of S_1, possibility for stimulated emission (red), and relaxation within vibrational and rotational states of S_0 form a 4-level lasing system. However, intersystem crossings from singlet (S_1) state to triplet state (T_1) occur with small probability, despite being spin-forbidden. Also the crossing back into the singlet state is an event with low probability. Meanwhile triplet absorption can take place. Over time, this reduces the overall occurrence of stimulated emission from S_1 and is a major reason why cw operation of dye lasers is hard to achieve.

The transitions between the states occur with very different probabilities. One can assign a lifetime (inverse of the probability to make this transition within a second) to each transition, such that some transitions are called fast, in the case of a high probability for the transition to take place, or slow in the other case. A transition between electronic states is most probable, when the overlap between both wave functions is highest, i.e. often with contribution of vibrational and rotational states. For incident light of a wavelength corresponding to the energy denoted by the transition in blue in Fig. 2.6, an electron can be excited into S_1 with some vibrational and rotational energy. There, a fast, non-radiative decay within S_1 will occur, before a photon might be emitted (red line) when going from S_1 to S_0, again into some vibration and rotational state above the absolute ground state. This transition can be stimulat-stimulated or spontaneous, the latter case being called fluorescence. It follows a fast relaxation to the ground state. This makes a four-level system, suited for lasing.

However, triplet states provide for non-radiative pathways and therefore strongly impact efficiency. An electron has to flip its spin in order to cross from the manifold of singlet states to the triplet state. This makes transitions between triplet and singlet states unlikely, and is called inter system crossing (ISC). Nevertheless, perturbations and collisions of molecules cause significant ISC rates. Molecules which go into triplet states cannot contribute to stimulated emission. Also, the transition from the triplet state back to the singlet is not very probable. The molecule therefore remains for some time in the triplet state, and is not available to contribute to lasing during that time. This is the main obstacle for cw operation of dye lasers. Triplet state absorption only leads to non-radiative relaxation and further increases the time the molecule stays in the triplet state.

To conquer this problem, triplet quenching methods, which accelerate the transition from T_1 into S_0, have been developed. For instance, an additional chemical agent can be introduced into the dye solution. This agent promotes transitions from T_1 to S_0 by collisions with dye molecules. In liquid laser dye solutions this is frequently done, for solid state dye lasers this might be less effective. Another method is to operate a dye laser at low repetition rate and simply wait for a complete return of the ensemble into S_0 or exchange participating molecules rapidly by making a fast flowing stream of dye molecules through the gain region of the laser.

Besides, triplet states are often more chemically reactive, which results in the transformation to another type of molecule, which cannot contribute to stimulated emission.

2.3.2 Lasing from Dyes

The described 4-level system exhibits homogenous broadening, meaning all molecules with an inverted population can contribute to stimulated emission at a certain wavelength within the resulting fluorescence spectrum. Vibrational and rotational states cause a relatively wide band. Consequently, this laser system is widely tunable (typical fluorescence linewidth 50…100 nm), and shows high gain. However, the short lifetime (a few ns) for a radiative transition from S_1 to S_0 also means that pump intensities have to be rather large in order to achieve sufficient inversion, see chapter 5 in [47].

Looking at the energy diagram, it also becomes apparent, that the pump photon energy is larger than the energy per emitted photon, because some excitation energy is lost during thermal relaxation in vibrational and rotational states. This so called Stokes shift significantly separates the peak of the absorption spectrum from the peak of the red-shifted emission spectrum around 50 nm for many dyes and enables efficient optical pumping.

Most dye lasers operate in the visible spectrum. The limit for shorter emission wavelengths is mainly determined by the photo stability of the dyes under the influence the pump light. At longer wavelengths a reduction of fluorescence quantum yield can often be observed, because more non-radiative transitions occur via thermal phonons from S_1 to S_0, simply because S_1 is much closer to S_0. These additional non-radiative pathways are limiting the choice of dyes beyond 1 µm and lead to low conversion efficiencies [48]. An advantageous side effect is that, also the lifetime of the transition from S_1 to S_0 is shortened. Hence ISC is less frequent, which improves the dye stability.

However, for longer emission wavelengths, there is another mechanism which negatively impacts the stability, because the lowest triplet state is closer to the ground state. Therefore it is more easily thermally populated, giving rise to chemical reactions and also reducing the fraction of excitable molecules. As a consequence, dyes with emission wavelength longer than 1.7 µm are very rare, according to [49].

2.3.3 Dyes on SOI Wafers

Any laser on silicon has to emit at a wavelength at which silicon is transparent, if silicon photonic structures such as arrayed waveguide gratings (AWGs), slot waveguides (light-matter interaction), filters, etc. are to be used. In [50], light emission using an organic cladding and silicon grating-based mirrors has been achieved for the first time on an SOI wafer, although at a wavelength of 625 nm. Choosing the right gain material decides the emission wavelength. Our work has been inspired by [51], who build the first IR solid state dye laser using polymer waveguides.

We use the same dye called IR26 [52], sum formula $C_{40}H_{30}Cl_2O_4S_2$, named 4-(7-(2-phenyl-4H-1-benzothiopyran-4-ylidene)-4-chloro-3,5-trimethylene-1,3,5-heptatrienyl)-2-phenyl-1-benzothiopyrylium perchlorate, which is depicted in the next figure.

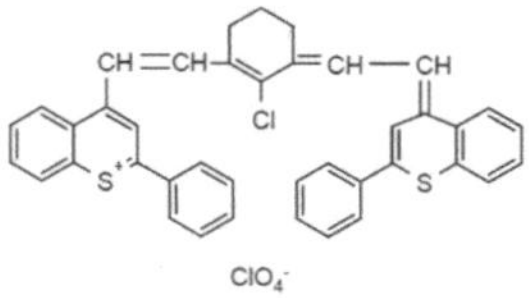

Fig. 2.7: Structure of dye IR26. Image source: Manufacturer Exciton Inc. [53]

Despite the often reported bleaching, IR26 has been developed especially for high photo stability at a pump wavelength of 1064 nm, the one we are also going to use.

Making integrated dye lasers on silicon reduces their cost immensely and enables their use in scenarios based on disposable chips. Their often criticized stability does not matter in this case, which is why dyes perfectly fit a number of applications in PICs made with SOI technology.

2.4 Electro-Optic Modulation on SOI

Modulating light with an electrical signal is useful in a range of applications. This work primarily focuses on application scenarios for SOH technology. This section serves to communicate the essential background for understanding the design of SOH modulators and the physical foundation. To put the accomplishments of SOH modulators in perspective, this section also contains a short summary about the underlying effects of pn-modulators on SOI. This allows for benchmarking of modulators fabricated on different platforms with respect to selected aspects.

Here, we only consider phase modulators, as these are the fundamental building blocks to make PICs even when using more complex modulation techniques. An amplitude modulation can easily be achieved by combining two phase modulators in a Mach-Zehnder interferometer.

Any phase shift φ will originate from a forced variation of the effective refractive index Δn_{eff} of the waveguide and depends on its length l and the wavelength λ of the guided light,

$$\varphi = -\frac{2\pi}{\lambda}\Delta n_{\mathrm{eff}}\, l\,. \tag{2.64}$$

The obvious question to be answered in the next two subsections is: How to induce a change of the effective refractive index of the waveguide?

2.4.1 Plasma Effect in SOI Structures

A frequently used mechanism to obtain modulation of light on SOI without the need for fabrication processes beyond the most established one from CMOS-fabrication based silicon photonics processes is the plasma effect.

Silicon's properties have been discussed in section 2.2.1 and the plasma dispersion effect introduced. To apply this effect to achieve modulation, a pn or pin-junction is created right in the center of a rib waveguide, see Fig. 2.8. When applying a voltage at the metal electrodes, the doped silicon at each side of the rip acts as a conductor. In this way, charges can be injected or depleted in the pn respectively pin junction. According to the model of Soref, see Eqs. (2.27) and (2.28), the equations are given here again for convenience,

$$\Delta\alpha_{\mathrm{Soref}} = 8.5\times10^{-22}\,\mathrm{cm}^2\Delta N_{\mathrm{e}} + 6.0\times10^{-22}\,\mathrm{cm}^2\Delta N_{\mathrm{h}}\,, \tag{2.65}$$

$$\Delta n_{\mathrm{Soref}} = -8.8\times10^{-22}\,\mathrm{cm}^3\Delta N_{\mathrm{e}} + 8.5\times10^{-18}(\mathrm{cm}^3\Delta N_{\mathrm{h}})^{0.8}\,, \tag{2.66}$$

an applied voltage will cause a change of the refractive index, with a marginal change in absorption. This makes this device a phase modulator. Applying an additional reverse bias allows faster operation, because the carriers are swept out, which avoids the speed limit otherwise set by their recombination time. Note that the depletion zone width changes during modulation, especially as many devices are modulated with voltages close to the reverse bias.

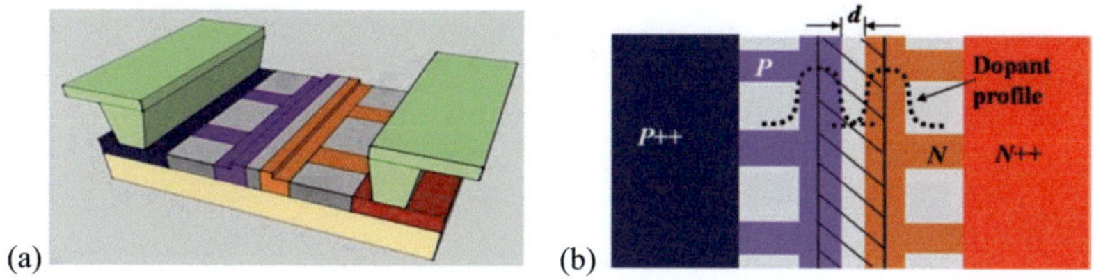

Fig. 2.8: Schematic view of a pin-modulator based on a rib waveguide on the SOI platform (a), and its doping scheme in top view (b). Color codes: buried oxide in yellow, metal electrodes in green, undoped silicon in grey, p-doped silicon in bluish colors, n-doped silicon in reddish colors. When applying a voltage at the metal electrodes, the doped silicon at each side of the rip acts as a conductor. In this way charges can be injected or depleted in the pn respectively pin junction, which is in the center of the rib waveguide. Light is guided along this rib. Changing the free carrier concentration changes the refractive index and therefore induces a phase shift. To reduce the attenuation of light by the doped silicon, it has been doped less (P,N) in the vicinity of the WG, than close to the metal electrodes (P++, N++). Image source: [J10]

2.4.2 Linear Electro-Optic Effect in SOH Structures

Using the linear-electro optic effect (Pockels effect) serves to induce a linear, pure phase shift with an external voltage. Widely used $LiNbO_3$ modulators employ this very effect in optical communications. These modulators are made of crystals with a $\chi^{(2)}$-nonlinearity. The same effect shall be used for modulators created with the SOH fabrication platform. By reducing the size of the devices, it is expected that similar electric fields can be achieved with much lower voltages. This improves the energy efficiency of these modulators. The aim is to make electrical amplifiers expendable when using SOH modulators.

Following p. 28 of [54] and rewriting Eq. (2.6), we can express the displacement field with the electric field $\mathbf{E} = (E_x, E_y, E_z)$ and the inverse of the dielectric permeability matrix $\varepsilon_{r,ij}$, the impermeability matrix η_{ij}, with $i, j \in \{x, y, z\}$ as

$$E_i = \sum_j \eta_{ij} D_j . \tag{2.67}$$

Assuming a lossless material, the matrix $\varepsilon_{r,ij}$ is symmetric and real, and so is η_{ij}, i.e., it can be reduced to six components. For the practical cases considered in this work it is sufficient to consider only the diagonal elements $\eta_{xx}, \eta_{yy}, \eta_{zz}$. To quantify the change of the impermeability $\Delta\eta_{ii}$ and the change of the related refractive index $\Delta\eta_{ii} = \Delta(1/n^2)_i$ for an applied electric field we use the electro-optic coefficients r_{ij}

$$\Delta\eta_{ii} = \Delta\left(\frac{1}{n^2}\right)_i = \sum_j r_{ij}E_j. \tag{2.68}$$

We assume the external field is applied in direction $\mathbf{e}_j$. Then the refractive index can be written (using the approximation from Appendix Eq. (7.2)) as

$$n_i + \Delta n_i = \sqrt{\frac{1}{\eta_{ii} + \Delta\eta_{ii}}} = \sqrt{\frac{1}{\eta_{ii} + r_{ij}E_j}} \simeq n_i - \frac{1}{2}n_i^3 r_{ij}E_j. \tag{2.69}$$

The change of the refractive index by small electric fields can be written in a linear approximation:

$$\Delta n_i \simeq -\frac{1}{2}n_i^3 r_{ij}E_j. \tag{2.70}$$

Remarkably any change is proportional to the 3rd power of the index of refraction. Hence a common figure of merit (FOM, here defined to include the unit m/V) to compare different EO materials is

$$\text{FOM}_{\chi^{(2)}} = n_i^3 r_{ij}. \tag{2.71}$$

The change in phase of quasi-TE-moded light at wavelength λ of for an electric field E_x applied in x-direction in a waveguide of length l containing a $\chi^{(2)}$-nonlinear material with refractive index n_x and interaction factor Γ_{Ex} can then be approximated using Eq. (2.34) as

$$\varphi = -\frac{2\pi}{\lambda}\Delta n_{\text{eff}}\, l \simeq \frac{2\pi}{\lambda} l\Gamma_{Ex}\frac{1}{2}n_x^3 r_{xx}E_x. \tag{2.72}$$

Materials Showing the Linear Electro-Optic Effect

We are using claddings from two distinct material configurations: *Nonlinear organic crystals* and *nonlinear chromophores in a polymer matrix*. The latter is sometimes also abbreviated and referred to as poled polymers, alluding to the method of creation of their $\chi^{(2)}$-nonlinearity. The refractive index is not the same for all directions. The next figure shows a coordinate system x,y,z in which the refractive index is indicated as an ellipsoid, such that one can determine which component of the E-field is subject to which refractive index. Note that the optical axis, also called the dielectrical axis, is defined as that axis, for which there is no birefringence. The refractive indices $n_x=n_1$, $n_y=n_2$, $n_z=n_3$, are experi-experienced by light with the electric field E in *x,y,z* direction, respectively. If indices are different for all directions then the material is called biaxial, if two refractive indices are the same the material is referred to as uniaxial, and if all are the same the material is isotropic.

While some materials such as $LiNbO_3$ and also poled polymers used in this work belong to the class of uniaxial materials, the organic crystal used here is biaxial. The crystal is orthorhombic. That means that its crystallographic axes *a,b,c* are parallel (respectively) to the axes *x,y,z* in the Cartesian coordinate system depicted in the next figure. Nevertheless, the unit cell's edge lengths are all different. The optical axis is not the same as the polar axis (crystallographic *c* axis).

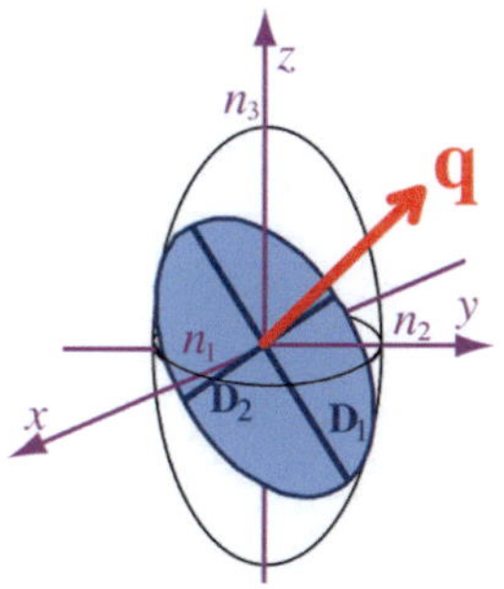

Fig. 2.9: Index ellipsoid with optical axis *q* marked in red. For light propagating in the direction of the optical axis, the experienced refractive index is the same for all polarizations. This means any two vectors, which are perpendicular to each other, perpendicular to the axis *q* and which are part of the index ellipsoid must have the same length ($D_1 = D_2$). In the case of an orthorhombic crystal, the crystallographic axes *a,b,c* are respectively parallel to the *x,y,z* axes. If refractive indices n_1, n_2, n_3 are all different and the material is called biaxial. If exactly two refractive indices are equal the material is called uniaxial and the index ellipsoid is also referred to as an indicatrix. If all refractive indices are equal, the material is called isotropic. Image source: Mojca Jazbinsek, Rainbow Photonics AG.

Organic Crystals as listed in the next table have been shown to exhibit a $\chi^{(2)}$-nonlinearity.

Integrating organic crystals on SOI allows using the EO effect for modulation. For applications the FOM and melting temperature are most important and determine the usefulness of organic crystals over other organic materials. However, practical questions on how to actually integrate these crystals on SOI limited any experimental activity to DC or AC measurements in the past.

Material	Chemical structure	Melting temp. (°C)	Stability at melting	Electro-optic coeff. *r*	Electro-optic figure-of-merit n^3r
DAST	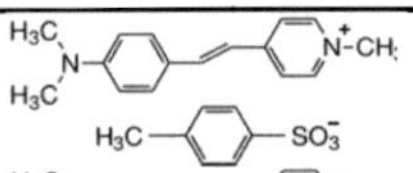	256	No (only short time)	48 pm/V	480 pm/V
DSTMS	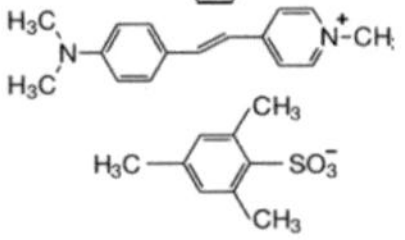	258	No (only short time)	48 pm/V	480 pm/V
OH1	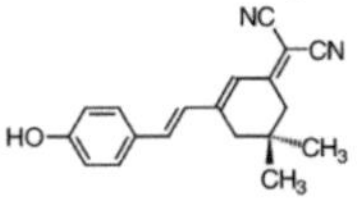	212	Yes	46 pm/V	470 pm/V
OH2	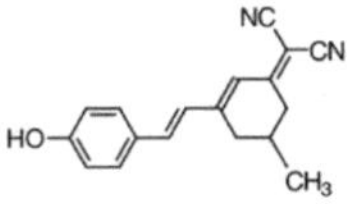	242	Yes	n.m. (expected > 60 pm/V)	n.m. (expected > 600 pm/V)
DAN	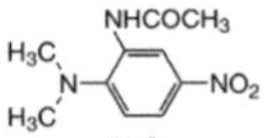	166	Yes	13 pm/V (@ 633 nm)	50 pm/V
BNA	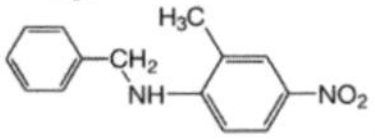	102	Yes	n.m. 24 pm/V calculated from d_{33}	n.m. 135 pm/V
DAT2	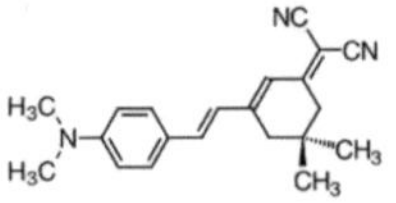	235	Yes	8 pm/V	30 pm/V

Table 2-1: List of organic crystals with $\chi^{(2)}$-nonlinearity. If a material is not stable at melting temperature then the material will decompose (chemical reaction) already at temperatures around or below the melting temperature. Not measured quantities are marked as n.m. Table source: Personal communication with Mojca Jazbinsek, Rainbow Photonics AG.

Due to the periodicity in repetition of a unit cell of an arrangement of chemically bound atoms in crystals one can identify three crystallographic directions, *a*, *b*, *c*. When growing a crystal, the energy for bonding atoms (and hence the probability to attach them to an already grown (seed crystal) differs according to the location of neighboring, bound atoms. Thus differently oriented planes emerge, which are determined by the speed of their growth. When growing a crystal in a thin film, as done in this work, the growth speed of a plane matters, because those planes which grow fastest will eventually form the thin film.

A number of growth rules, illustrated in the next figure, can be observed:

- The crystallographic axis *b* is always perpendicular to the thin film.
- Hence the thinnest distance is in direction of *b*, which means the directions of the crystallographic axes *a*, *c* are fixed with respect to the seed crystal.
- When intending to apply an electric field across a waveguide and along the polar axis *c* of the material (to make use of the r_{33} coefficient to obtain an EO effect), the crystal has to be grown with its direction *a* in parallel to the WG axis.

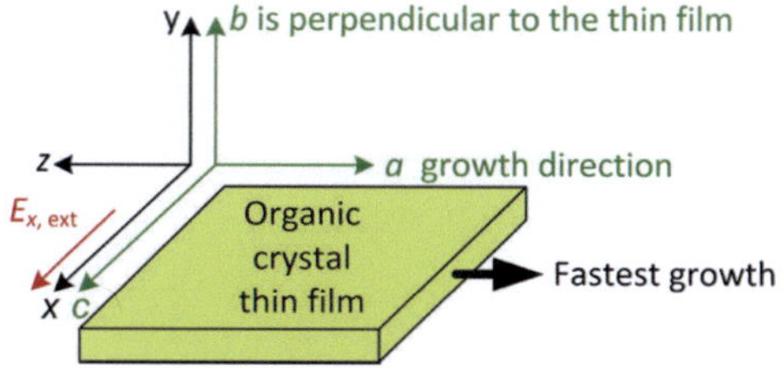

Fig. 2.10: Schematic view of thin film growth or organic crystals. *x,y,z* coordinate system for waveguide orientation, shown in black. The *a,b,c* coordinate system denotes the crystallographic axes of the organic crystal, shown in green. Its growth is confined to make a thin film in *y* direction. Crystallographic axis *b* will align to *y* during growth. The crystal plan normal to the crystallographic axis *a* growth fastest. If the direction of an applied, external electric field (red) coincides with the crystallographic axis *c* the (usually) largest linear electro optic coefficient can be used, e.g., to obtain a phase modulation.

In this work, successful deposition of the organic crystal BNA on SOI waveguides is shown, and proven by making high-speed data modulation.

Chromophores Hosted in a Polymer Matrix present another material option for obtaining the Pockels effect. Chromophores have a conjugated pi-bond system. By choosing molecules, which are especially easy to polarize (molecular hyperpolarizability [55]) and for which this occurs within an asymmetric potential, materials with the possibility for a high macroscopic $\chi^{(2)}$-nonlinearity can
be created.

To transform the microscopic hyperpolarizability into an EO effect accessible at the macroscopic scale, a number of further conditions have to be met: The molecules have to be aligned along the same direction and with the same orientation. One way to reach this goal is hosting the chromophores in a polymer matrix. To change the initial random orientation, the material needs to be heated close to its glass transition temperature allowing the rotation of molecules. A temporally constant external electric field is applied which exerts a force on each molecule and thus aligning it. If the temperature is reduced to operating conditions the molecules are frozen in their new positions and a macroscopic, permanent uniaxial nonlinearity results, even without any external field. For that procedure to work, the material has to be non-conductive.

A high refractive index is of advantage for the FOM, as are photo stability and a high glass transition temperature. It has been found that molecule end groups also significantly influence the final arrangement of the molecules, due to interaction between the chromophores. This interaction is of particular influence, when the amount of polymer is reduced or the host matrix even completely avoided.

The most stable chromophores used in this work are in the commercial material M3 from GigOptix Inc. (specified as $r_{33} = 70$ pm/V). The best nonlinearity reached in-house is $r_{33} = 190$ pm/V with DLD164. An overview about other interesting materials is given in [56].

Note that these materials are specifically designed for EO modulation. Investigating them in SOI structures with strip-loaded slot WGS is of particular advantage, as break-through currents during poling are naturally limited by the resistance of the thin silicon slabs. A break through therefore remains local, and will not be supplied with enough electrical power to heat up and destroy the material beyond that point.

In Summary, it can be said that both cladding options, organic crystals and chromophores aligned by poling, have their specific advantages and disadvantages. Organic crystals promise stability at high temperatures, strong light intensity. Their single crystalline growth on SOI wafers requires further process development.

In contrast, chromophores hosted in a polymer matrix currently deliver higher $\chi^{(2)}$-nonlinearities at reliable fabrication yield from a well-mastered poling process. This poling has to be done for every waveguide and varies in efficiency. Polymer based modulators are thought to be less stable during long-term operation at elevated temperatures, despite Telecordia certification of devices with selected materials.

2.5 SOH Travelling Wave Modulator

The propagation of the optical wave and the RF signal as well as their interaction need to be controlled. To optimally use the capabilities of nonlinear organic claddings for SOH modulators, strip-loaded slot WGs are used, as introduced in Subsection 2.2.2.

We first describe the modulator properties which are in focus for any application. A model is made for the modulator's frequency response according to the proposed implementation. This model is subsequently refined, as required by different limiting effects, which are discussed one by one. Design conclusions are drawn from this model also taking technical constraints into account.

2.5.1 Modulator Properties

The following list of properties is used to optimize modulators (any modulator, not just SOH modulators). A balance has to be found for a number of target properties, as some of these require trade-offs. Among the most application relevant properties are:

- RF-bandwidth (BW) of the modulator depends on its frequency response, which is deciding the modulators maximum operating speed (e.g. symbol rate). There is no point in optimizing a modulator for bandwidth much beyond the bandwidth of available electronic drivers for digital applications.

- π-voltage, the voltage needed to induce a π-phase shift, $\varphi = \pi$. This property is most relevant, because a modulator needs driving electronics. Reducing the required voltage reduces the demands for the electronics and the energy consumption. Note that the power of a signal scales with the square of the used voltage. Hence a design goal must be to keep the π-voltage small.
- Insertion loss (IL), i.e., the fraction of light lost in the device, which also strongly impacts the optical signal-to-noise ratio (OSNR) and thus the quality of the channel/subchannel.
- Extinction ratio (ER). When using interferometric structures such as MZIs to make MZMs or IQ modulators, loss in each arm has to be well controlled to achieve a good contrast between constructive and destructive interference in order to differentiate between different symbols. The extinction ratio ER_{DC} can be measured for an MZI without modulation by comparing the optical output power in case of maximum transmission $P_{\text{out, max}}$ and minimum transmission $P_{\text{out, min}}$ (apply the corresponding DC bias voltages) by adjusting the bias points accordingly.

$$\text{ER}_{\text{DC}} = \frac{P_{\text{out, max}}}{P_{\text{out, min}}} \tag{2.73}$$

  The ER during modulation, which naturally depends on the applied voltage compared to the π-voltage, is defined for a 2-level intensity modulation (on-off-keying, OOK) as the ratio of the optical power at the high level ("1") $P_{\text{out, high}}$ and low level ("0") $P_{\text{out, low}}$.

$$\text{ER}_{\text{MOD}} = \frac{P_{\text{out, high}}}{P_{\text{out, low}}} \tag{2.74}$$

  The optical power during modulation can for instance be measured with an optical digital communication analyzer (DCA).
- Linearity of the phase shift as a function of electrical drive voltage. This property becomes relevant when using advanced modulation formats. This is assured in principle by the Pockels effect for SOH modulators.

In addition, practical considerations impose further constraints, such as the desire for wavelength transparency, i.e. being able to use the modulator for a large range of different wavelengths, maybe even at the same time (frequency comb line generation, pulse carving, etc.). This work therefore concentrates on an optically non-resonant design by employing Mach-Zehnder interferometers (MZI), which combine phase modulators.

2.5.2 Modulator Modeling

The modeling of the phase modulator section in this work follows closely the procedure of Witzens et al. [57] and [J14]; in a slightly modified way to take a more advanced CMOS metal stack into account [58].

The length of around 1 mm of a modulator is in the same order of magnitude as the shortest contributing RF wavelength corresponding to an RF frequency of 30 GHz. Hence the propagation of the RF wave along the same interaction area as the optical signal should be optimized in a travelling wave (TW) approach.

Implementation of an RF travelling wave modulator

The previously introduced strip-loaded slot waveguide guarantees a high overlap between optical quasi TE-mode (dominant electrical field component in x-direction) and electrical field right in the slot of the WG. Both electrical fields are depicted in Fig. 2.11(a,b). The strip-loaded slot WG determines the light propagation, its doped Si slabs allow bringing an electrical signal to the Si rails, but the actual RF wave needs a waveguide of its own in a travelling wave design.

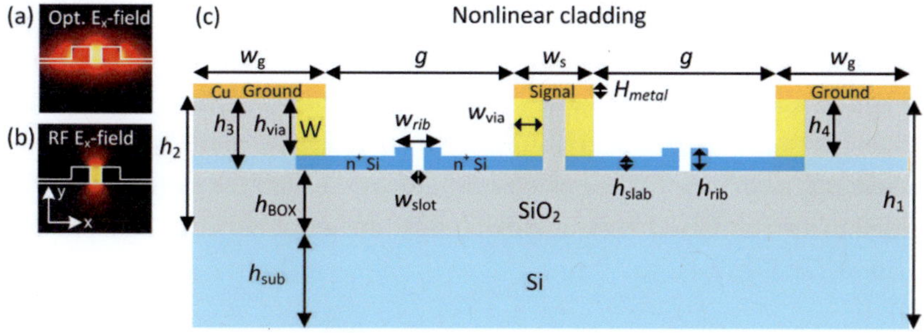

Fig. 2.11: Cross sections of strip-loaded slot WG and the SOH travelling wave MZM modulator. (a) Dominant E_x-component of the optical quasi TE mode. (b) Dominant E_x component of the applied RF field assuming well-conducting Si slabs. Note that both E-fields (optical and RF) are strong within the slot of the WG, i.e., exactly where the nonlinear material is placed. (c) Schematic cross section the SOH travelling wave MZM modulator, covered with nonlinear material.

We use a coplanar waveguide (CPW) in ground-signal-ground (GSG) electrode configuration. The complex propagation constant γ of the RF wave in this CPW can be written as the sum of the real-valued propagation constant β_{RF} and a real-valued attenuation term $\alpha_{\mathrm{RF,amp}}$:

$$\gamma = \alpha_{\mathrm{RF,amp}} + \mathrm{j}\beta_{\mathrm{RF}} \tag{2.75}$$

This RF WG can supply two strip-loaded slot WG-based phase modulators. The electrodes are made of copper, which is chosen for its high conductivity, and are connected with tungsten vias to doped Si slabs, as illustrated in Fig. 2.11(c).

The Eq. (2.72) is repeated here.

$$\varphi = -\frac{2\pi}{\lambda}\Delta n_{\mathrm{eff}}\, l \simeq \frac{2\pi}{\lambda} l \Gamma_{Ex} \frac{1}{2} n_x^3 r_{xx} E_x . \tag{2.76}$$

Assuming a parallel plate capacitor-like electric field in the slot $E_x = V_{\pi,\mathrm{PM}} / w_{\mathrm{slot}}$ the π-voltage ($\varphi = \pi$) $V_{\pi,\mathrm{PM}}(f)$ for one phase modulator at low RF frequencies (DC case) can be expressed as $V_{\pi,\mathrm{PM\ at\ DC}}$

$$\left|V_{\pi,\mathrm{PM\ at\ DC}}\right| = \frac{\lambda w_{\mathrm{slot}}}{l \Gamma_{Ex} n_x^3 r_{xx}} . \tag{2.77}$$

The proper selection of the type of RF WG is driven by the intended application of the modulator. Combining two phase modulators in an MZI allows making a Mach-Zehnder modulator (MZM), which in turn enables amplitude modulation. Operating a MZM in push-pull mode (as done with all MZMs used in this work) means creating a positive phase shift with the same magnitude in one arm and a negative phase shift in the other by just applying a voltage at the signal electrode. The entire structure is filled with a highly efficient $\chi^{(2)}$-nonlinear material. The $\chi^{(2)}$-nonlinearity must have the same orientation in both slots. Thus, this MZM only needs half the voltage to go from minimum to maximum transmission (make an advance of π in its transfer function, see Fig. 2.4).

$$V_{\pi,\mathrm{MZM}}(f) = \frac{1}{2} V_{\pi,\mathrm{PM}}(f) . \tag{2.78}$$

To describe the modulator's RF performance we refer to its EO frequency response $S_{21,\mathrm{EO}}(\omega_{\mathrm{RF}})$. Using the Pockels effect means that any change of voltage (small signal) applied across the slot is converted linearly to a change in power of the MZM, when biased at the quadrature point. As the Pockels effect is not dependent on the RF frequency (denoted as f from hereon, $\omega_{\mathrm{RF}} = 2\pi f$), the bandwidth of the modulator is assumed to be determined by the traveling wave electrode design[13]. Neglecting any reflections of the propagating RF wave, we define the contribution of the traveling wave electrodes $S_{21,\mathrm{V}}(f)$ to the overall frequency response of the modulator $S_{21,\mathrm{EO}}(\omega_{\mathrm{RF}})$ as the ratio of the voltage $|V_{\mathrm{RF}}(f)| = |V_{\pi,\,\mathrm{MZM}}(f)|$ applied at the GSG electrodes input and the effective voltage $|V_{\mathrm{slot}}| = |V_{\pi,\mathrm{PM\ at\ DC}}|$ actually applied across the slot (factor ½, because the MZM is in push-pull operation, see Eq. (2.78)).

$$|S_{21,\mathrm{V}}(f)| = \left|\frac{V_{\mathrm{slot}}}{V_{\mathrm{RF}}(f)}\right| = \frac{1}{2}\frac{\dfrac{\lambda w_{\mathrm{slot}}}{l\Gamma_{Ex} n_x^3 r_{xx}}}{|V_{\pi,\,\mathrm{MZM}}(f)|} \propto |S_{21,\mathrm{EO}}(f)|. \tag{2.79}$$

Design trade-off

Ideally the refractive index n_x of the nonlinear cladding would be chosen to balance the FOM (improving with n_x^3) and the interaction factor Γ_{Ex} (decreasing with higher n) of light in the slot. However, the choice of this cladding is currently determined by availability and reliability of the material.

Design trade-off

The confinement of light to the slot (and thus the achievable phase shift) improves when reducing the slot width w_{slot}. However, scattering losses increase strongly for smaller widths, due to fabrication induced roughness of the WG sidewalls. In this work we therefore had to choose slots with a width of 120…140 nm.

The numerical field analysis to find the modes of the optical signal, and to investigate the propagation of the RF wave, was carried out with the help of the CST Microwave Studio software package. It is nevertheless instructive to look at an equivalent circuit model [59] of the SOH modulator in order to understand the performance of the device.

---

13 Nevertheless for some materials, such as chromophores in a polymer matrix, a low-frequency dependence (< 1 GHz, not part of the Pockels effect) of the effective refractive index on the applied voltage has been observed, which is neglected in the considerations presented here.

Equivalent circuit model approach

The modeling has to address the following four issues:

- Walk-off between optical and electrical wave
- RC limitation of the strip-loaded slot
- Impedance mismatch
- Travelling wave losses

We start from a very rough model, a resistance-capacitance (RC) element, and extend it to a transmission line based model to understand the behavior of the modulator. While basic circuit theory cannot model the RF behavior of the devices (the line length is larger than the RF wavelength), transmission line theory allows modeling the device as a series of infinitesimally short lumped elements, i.e., differential elements (marked with ′). A more detailed analysis is given by Witzens et al. [57]. To include the influence of a CMOS metal stack we adapt the equivalent circuit model proposed by Hui et al. in [58].

2.5.3 Walk-Off Between Optical and Electrical Wave

For optimal modulation the optical and electrical wave should travel at the same speed. The group velocity of light[14] and the group velocity[15] of the RF signal are to be compared. Using CST Microwave Studio or COMSOL the effective group refractive index of the strip-loaded slot WG was determined to be $n_{g,\,opt} = 3.1$ and the group effective refractive index of the RF wave at 40 ± 10 GHz is $n_{g,\,RF} = 2.7$ ($n_{RF} = 3.0$). Considering the typical length of the modulator of 0.5…1.5 mm, the walk-off effect is thus neglected when modeling the overall modulator behavior.

2.5.4 RC Limitation of the Strip-Loaded Slot

While the RF signal travels along the coplanar waveguide (CPW), it is brought to the slots with the help of W-vias and the conductive silicon slabs. In fact, the tungsten vias are much more conductive than the Si slabs (which are thin

---

[14] Considering for instance intensity modulation, information is to be encoded to make light pulses, which travel with group velocity.

[15] The refractive index change of the nonlinear cladding follows the RF voltage, which is dependent on the group velocity of the RF signal.

strips), such that only the slab's resistance is modeled (denoted with G_{Si}^{-1}) in the following consideration. The slot itself can be modeled as a capacitor with C_{slot} which needs to be charged via the ohmic resistance of the silicon strips (see Fig. 2.12(a), consider only the slot capacitance and Si slab resistance). This series connection of resistance and capacitance has an RC constant.

Charging and discharging the slot takes time, characterized by the time constant τ with corresponding 3 dB bandwidth f_{RC}, which can limit the modulation speed:

$$f_{\mathrm{RC}} = \frac{G_{\mathrm{Si}}}{2\pi C_{\mathrm{slot}}} \text{ and } \tau = \frac{C_{\mathrm{slot}}}{G_{\mathrm{Si}}}. \tag{2.80}$$

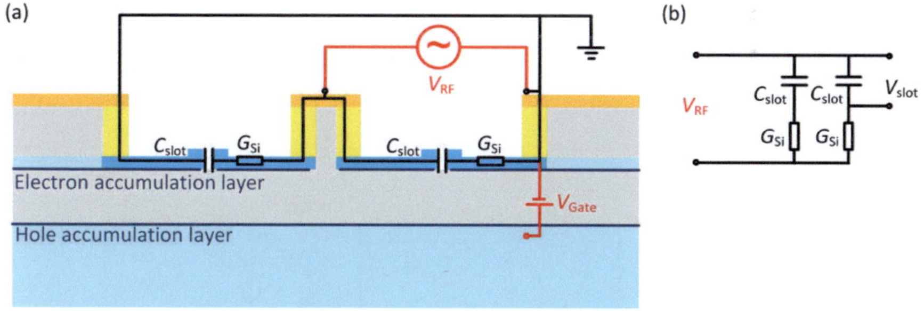

Fig. 2.12: Equivalent circuit modeling the slot capacitance C_{slot} and Si slab conductance G_{Si} as a lumped element. (a) Cross section of the SOH MZM, with RC elements marked. The conductance of the left and right slab of each WG has been merged into one symbol G_{Si}. A DC voltage V_{Gate} can be applied to the substrate to create an electron accumulation layer (similar to the channel in a FET) to increase the conductivity in the Si slabs connecting the CPW with the Si rails on both sides of the slot. (b) Equivalent circuit of the RC lumped element.

Design choice

The capacitance could be reduced by making the slot wider. However, this approach would result in a larger π-voltage, see Eq. (2.77), and is therefore disregarded. Note that the slot capacitance is fixed by the geometry in SOH modulators.

Design trade-off

The implant concentration of the Si slabs could be increased to reduce their resistance. However, according to Eq. (2.27) this would also increase insertion loss. We use the lowest possible doping to reach a reasonable bandwidth. The trade-off between IL and BW is not further discussed, as fabrication is not pre-predictable enough (e.g., the doping profile, and resulting conductivity; the influence of surface states). We therefore rely on the observation in [J14] and use the solution proposed in the patents [P1], [P2]. A positive voltage V_{Gate} can be applied to the substrate, see Fig. 2.12(a), which creates an electron accumulation layer (very similar to the channel in a FET) in the Si slabs. This accumulation layer makes the Si slabs more conductive without introducing excessive optical losses. The advantage of carriers introduced this way over obtaining free carriers from implanting ions is discussed in more detail in [33].

The resulting RC time constant and the corresponding 3 dB bandwidth can be calculated. We measured the capacitance of a typical SOH phase modulator ($C'_{\text{slot}} = 400$ fF/mm, measurement provided by R. Palmer, W. Heni), and the resistance over the entire structure (of a device, in which the slot was not created and is "filled" with Si). Fig. 2.13(a) depicts the resistance vs. the applied gate voltage. Fig. 2.13(b) shows the resulting 3 dB frequency f_{RC}.

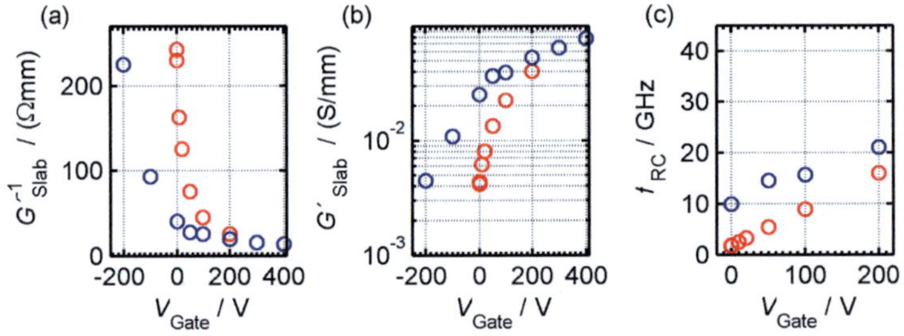

Fig. 2.13: Measured resistance of SOH phase modulator vs. DC voltage V_{Gate} applied to the substrate for two devices from the same wafer. The measurements represented in blue will be used further, as they match later results. The measurements depicted in red are included to indicate problems at fabrication. (a) Measured differential resistance, assumed to originate almost entirely from the Si slabs. (b) Measurements expressed as conductance. (c) Resulting RC constant calculated from measured C_{slot}.

Assuming such a lumped equivalent circuit, the frequency response is

$$\left|S_{21,\text{V}}(f)\right| = \left|\frac{V_{\text{slot}}}{V_{\text{RF}}(f)}\right| = \left|\frac{1}{1 + \text{j}2\pi f G^{-1} C}\right|. \tag{2.81}$$

The resulting π-voltage of the MZM can be written:

$$\left|V_{\pi,\,\mathrm{MZM}}(f)\right| = \frac{1}{2}\frac{\lambda w_{\mathrm{slot}}}{l\Gamma_{\chi^{(2)}\mathrm{reg}}\, n^3_{x,\chi^{(2)}\mathrm{reg}}\, r_{xx}}\sqrt{1+(f/f_{\mathrm{RC}})^2} \tag{2.82}$$

Relying on the previous measurements, the π-voltage's dependence on frequency is shown in Fig. 2.14.

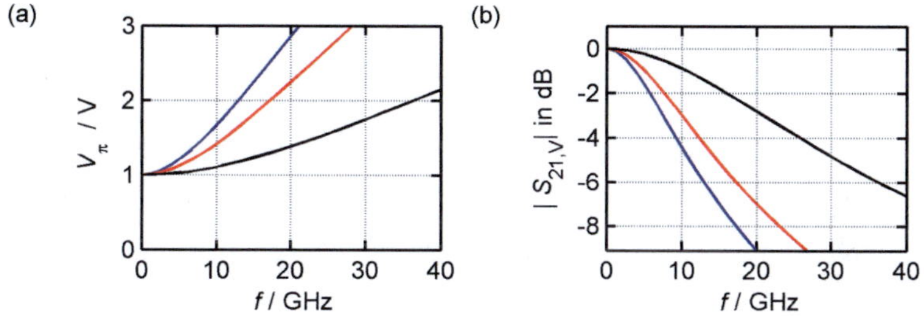

Fig. 2.14: (a) Calculated frequency dependence of π-voltage (normalized to 1 V at DC) and of the resulting (b) $S_{21,\mathrm{V}}$ parameter. Blue, red, black curves depict a f_{RC} of 7.5 GHz (measurement with contact resistance, $V_{\mathrm{Gate}} = 0$ V), 10 GHz (measurement without contact resistance, $V_{\mathrm{Gate}} = 0$ V) and 21 GHz ($V_{\mathrm{Gate}} = 200$ V), respectively.

2.5.5 Impedance Mismatch

An impedance mismatch between the circuit which generates the electrical drive signal (generator) having an impedance Z_{gen} and the SOH modulator with a characteristic impedance Z_0 will lead to a reflection of the incoming electrical signal. This reflection is to be avoided for efficiency reasons, and also to avoid "echoes" of the data stream. Another reflection to avoid can occur at the interface of the electrical output of the modulator and its termination. The SOH modulator is usually terminated with a 50 Ω load, called impedance Z_{load}.

The reflection between device and load can be calculated (voltage ratio of the reflected V_{r} and incident wave V_{i}) as an amplitude reflection factor

$$\Gamma_{\mathrm{load}} = \frac{V_{\mathrm{r}}}{V_{\mathrm{i}}} = \frac{Z_{\mathrm{load}} - Z_0}{Z_{\mathrm{load}} + Z_0}\,. \tag{2.83}$$

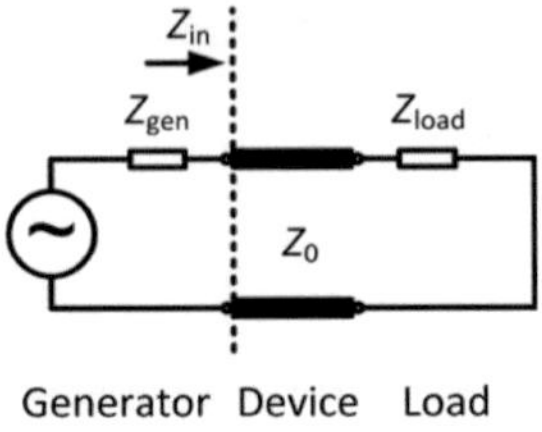

Fig. 2.15: Equivalent circuit representing the concatenation of driver circuit (generator) with impedance Z_{gen}, SOH modulator with impedance Z_0, and terminating load with impedance Z_{load}. The device with load has an input impedance Z_{in}.

To also determine the reflection between loaded modulator and the signal generator, the input impedance Z_{in} of the modulator (length l, real valued propagation constant β_{RF}) with load is first calculated as

$$Z_{in} = Z_0 \frac{Z_{\mathrm{load}} + \mathrm{j}Z_0 \tan\left(\beta_{\mathrm{RF}} l\right)}{Z_0 + \mathrm{j}Z_{\mathrm{load}} \tan\left(\beta_{\mathrm{RF}} l\right)}. \tag{2.84}$$

Thus the amplitude reflection factor of the terminated device can be written as

$$\Gamma_{\mathrm{term.dev}} = \frac{Z_{\mathrm{in}} - Z_{\mathrm{gen}}}{Z_{\mathrm{gen}} + Z_{\mathrm{in}}}. \tag{2.85}$$

When employing on-chip drivers, these can certainly be designed for an impedance other than the standard line impedance of 50 Ω. In this work external drivers have been used. In order to avoid reflections and wasting electrical signal energy, the SOH modulators have been designed to reach an impedance of around 50 Ω. For that, the widths of the CPW electrodes and their spacing have been optimized using CST Microwave Studio. The experimentally derived characteristic impedance (absolute value) of a typical SOH MZM is shown in the next figure. Its impedance is sufficiently close to the standard line impedance, except for lower frequencies, which is an advantage for the overall frequency response, as it compensates in part the characteristic of the RC element.

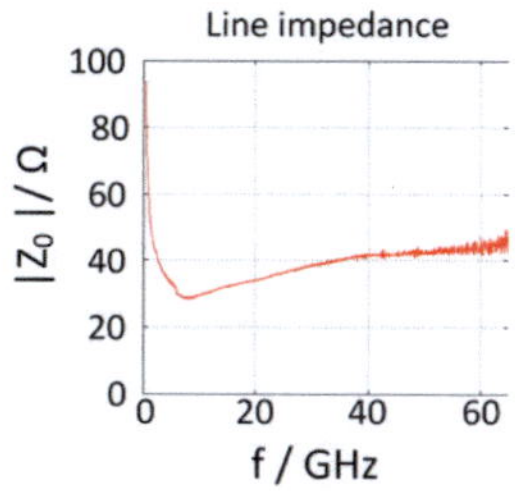

Fig. 2.16: Absolute characteristic line impedance of a typical MZM SOH modulator, measured with a small RF signal in the quadrature operating point.

Design trade-off

The spacing between CPW electrodes g and the width w_S of the signal electrode could be further reduced to match the device's impedance better to the standard line impedance. However, their current dimensions are already close to limits imposed by the CMOS fabrication processes, and could therefore not be further optimized.

2.5.6 Travelling Wave Losses

Microwave losses depend on the frequency of the travelling wave. The skin effect is a major contributor for CPWs and can be quantified with the skin depth for good conductors (μ is the absolute magnetic permeability of the conductor, σ its conductivity) such as copper using (p. 19 in [27])

$$\delta_{\text{skin}} = \sqrt{\frac{1}{\pi f \mu \sigma}}. \tag{2.86}$$

For thick electrodes the attenuation of the guided RF wave is inversely proportional to the skin depth, because the effective cross section used for conduction is reduced with increasing frequency. In our implementation the thickness of the Cu electrodes is fixed by the CMOS based fabrication process at 0.5 µm. The skin depth for Cu at 10 GHz is 0.65 µm.

The higher the frequency, the higher is the propagation loss, which can be characterized with a power attenuation constant $\alpha_P(f)$.

$$\frac{\mathrm{d}P_{\text{RF}}(f)}{\mathrm{d}z} = -\alpha_P(f) P_{\text{RF}}(f) \tag{2.87}$$

This results in an effective length l_{eff} [57]. The description of the effect of l_{eff} with a power attenuation constant on the π-voltage (amplitude) requires the fac-factor ½ in Eq. (2.88).

$$l_{\text{eff}} = \int_0^l \exp(-\alpha_{\text{P}}(f) z/2)\,\text{d}z \tag{2.88}$$

The π-voltage of the modulator is inversely proportional to the effective length.

$$\left|V_{\pi,\,\text{MZM}}(f)\right| = \frac{1}{2}\frac{\lambda w_{\text{slot}}}{l_{\text{eff}}(f)\Gamma_{\chi^{(2)}\text{reg}} n^3_{x,\chi^{(2)}\text{reg}} r_{xx}}\sqrt{1+(f/f_{\text{RC}})^2} \tag{2.89}$$

Measured RF losses of a typical 1 mm long SOH modulator are shown in Fig. 2.17. The frequency dependence of the *power attenuation coefficient* appears to be of almost linear nature (with an offset of 1 dB/mm).

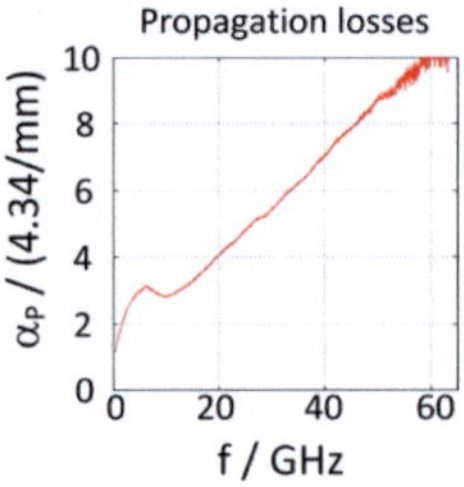

Fig. 2.17: Propagation losses expressed as power attenuation coefficient of a typical SOH modulator with grounded substrate. Image source: Courtesy of R. Palmer, KIT. As α_{P} is part of an exponent to calculate the attenuation, expressing the attenuation in dB/mm means a multiplication of α_{P} with 4.344, i.e., reading the vertical axis of this plot gives directly the attenuation in dB/mm.

To understand the frequency response of the SOH modulator, and in particular the RC limitation and RF propagation losses, a more detailed model is needed. Witzens et al. [57] model the SOH modulator with a transmission line (TL). For this, an infinitesimal short lumped-element circuit with differential elements series resistance R', series inductance L', shunt capacitance C', and shunt conductance G' are defined.

$$C = \int_0^l C'\text{d}z\ ,\ G = \int_0^l G'\text{d}z\ ,\ R = \int_0^l R'\text{d}z\ ,\ L = \int_0^l L'\text{d}z \tag{2.90}$$

Fig. 2.18(a) shows a revised model including the shunt capacitance C'_{metal} of the CPW. Moreover, the series resistance R'_{TL} and series inductance L'_{TL} of the TL complete the infinitesimal short lumped-element circuit shown in Fig. 2.18(b). For calculations the series RC connection is transformed into a parallel RC connection in Fig. 2.18(c) using the definition of the quality factor Q_{S} of a series RC connection,

$$Q_{\mathrm{S}} = \frac{1}{\omega G'^{-1} C'}, \tag{2.91}$$

$$G'^{-1}_{\mathrm{TL}} = \frac{1}{2}\left(1 + Q_{\mathrm{S}}^2\right) Z'_{\mathrm{Si}} = \left(1 + \frac{1}{\omega^2 G'^{-2}_{\mathrm{Si}} C'^2_{\mathrm{slot}}}\right) G'^{-1}_{\mathrm{Si}}, \tag{2.92}$$

$$C'_{\mathrm{TL}} = \frac{2C'_{\mathrm{slot}}}{\left(1 + Q_{\mathrm{S}}^{-2}\right)} + C'_{\mathrm{metal}} = \frac{2C'_{\mathrm{slot}}}{1 + \dfrac{1}{\omega^2 G'^{-2}_{\mathrm{Si}} C'^2_{\mathrm{slot}}}} + C'_{\mathrm{metal}}\,. \tag{2.93}$$

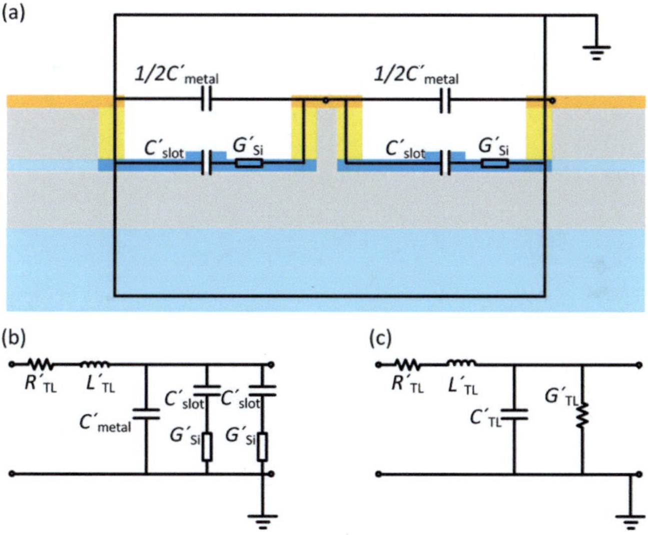

Fig. 2.18: Equivalent circuit model with the slot capacitance C'_{slot} and Si slab conductance G'_{Si} in a transmission line. (a) Equivalent circuit components corresponding to the cross section are shown. (b) The RF wave is modeled with a series of infinitesimal short lumped-element circuits, as shown here. The transmission line (TL) experiences a series resistance R_{TL} and series inductance L'_{TL}, as well as a shunt capacitance C'_{metal} and the RC elements of the strip-loaded slot WG. (c) To apply the telegrapher's equations, the circuit of (b) is transformed to this standard form with only on shunt capacitance C'_{TL} and one shunt conductance G'_{TL}.

The characteristic impedance of a transmission line is derived from the Telegra-Telegrapher's equations and is in general

$$Z_0 = \sqrt{\frac{R'+\mathrm{j}\omega L'}{G'+\mathrm{j}\omega C'}}. \tag{2.94}$$

Using this equation and assuming the CPW's series resistance can be neglected, because the CPW is made of Cu, the characteristic impedance to the equivalent circuit in Fig. 2.12(c) is

$$\begin{aligned} Z_0 &= \sqrt{\frac{L'_{\mathrm{TL}}}{C'_{\mathrm{TL}}+G'_{\mathrm{TL}}/\mathrm{j}\omega}} = \sqrt{\frac{L'_{\mathrm{TL}}}{C'_{\mathrm{metal}}+2C'_{\mathrm{slot}}\frac{1}{1+\mathrm{j}G'^{-1}_{\mathrm{Si}}C'_{\mathrm{slot}}}}} \\ &= \sqrt{\frac{L'_{\mathrm{TL}}}{C'_{\mathrm{metal}}+2C'_{\mathrm{slot}}\frac{1}{1+\mathrm{j}f/f_{\mathrm{RC}}}}} \end{aligned}. \tag{2.95}$$

The complex propagation constant of a transmission line is in general

$$\gamma = \alpha_{\mathrm{amp}} + \mathrm{j}\beta_{\mathrm{RF}} = \sqrt{(R'+\mathrm{j}\omega L')(G'+\mathrm{j}\omega C')}. \tag{2.96}$$

For a low-loss line, α_{amp} can be approximated (p. 79 in [27]) as

$$\alpha_{\mathrm{amp}} \simeq \frac{1}{2}\left(R'\sqrt{\frac{C'}{L'}} + G'\sqrt{\frac{L'}{C'}}\right) = \frac{1}{2}\left(\frac{R'}{Z_0} + G'Z_0\right), \tag{2.97}$$

$$\text{with } Z_0 \simeq \sqrt{\frac{L'}{C'}}. \tag{2.98}$$

The last expression is only exact in the absence of losses and should be a real quantity. Witzens et al. [57] obtain a relation for the transmission losses. The amplitude attenuation coefficient α_{amp} for the equivalent circuit in Fig. 2.12(c) is:

$$2\alpha_{\mathrm{amp}} = \alpha_{\mathrm{P}} \simeq \left(\underbrace{\frac{R'}{Z_0}}_{\simeq 0} + G'Z_0\right) \simeq \Re\left\{\frac{2\mathrm{j}\omega C'_{\mathrm{slot}}}{1+\mathrm{j}f/f_{RC}}\right\}Z_0 = \frac{(2\pi f)^2 C'^2_{\mathrm{slot}} G'^{-1}_{\mathrm{Si}}}{1+f^2/f^2_{RC}} Z_0 \tag{2.99}$$

Witzens et al. [57] argue that it is useful to normalize these losses with Z_0^{-1}, because V^2/Z_0 is the power transported over the transmission line.

The resulting power attenuation coefficient can be written with a constant a_{Wi} (which contains a factor of 2 for amplitude to power conversion, Z_0 is elim- eliminated; this constant is assumed to be not dependent on frequency) and the bandwidth f_{RC} as

$$\alpha_{\mathrm{P}} \simeq a_{\mathrm{Wi}} \frac{f^2}{1+\left(\frac{f}{f_{\mathrm{RC}}}\right)^2}. \tag{2.100}$$

Based on the model of the power attenuation coefficient by Witzens et al. [57] and assuming an RC-limited BW of 10 GHz, the attenuation affects the frequency response of the modulator, as shown in Fig. 2.19. To plot these curves, the constant a_{Wi} is chosen such that the attenuation corresponds to (1, 3, 7) dB/mm at 40 GHz, respectively plotted in blue, red and black. Note that beyond 10 GHz, the attenuation change reduces, because the RC element ceases to contribute to the attenuation; i.e., the slot capacitance is hardly loaded anymore.

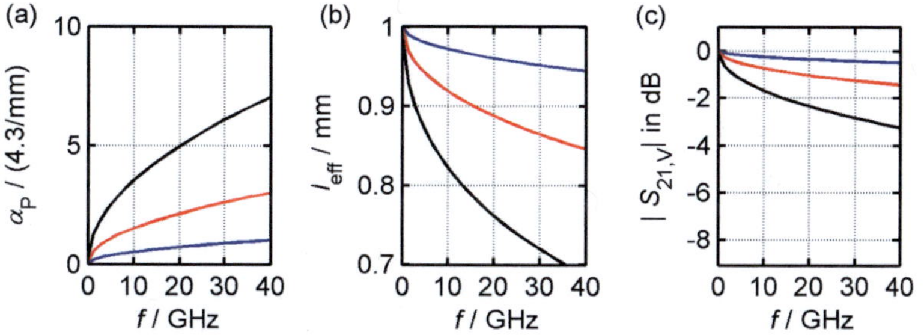

Fig. 2.19: Frequency dependence of power attenuation coefficient, effective length and the contribution to the modulator's response; according to the transmission line based model of [57]. The blue, red and black curves correspond to an attenuation of (1, 3, 7) dB/mm at 40 GHz, respectively. (a) Frequency dependence of the power attenuation coefficient α_{P} of the RF wave, assuming an RC-limited BW of 10 GHz. (b) Frequency dependence of the related effective length l_{eff}. (c) The resulting absolute value of the EO $S_{21,\mathrm{V}}$ frequency response.

Combining the RC-limit and the transmission loss related frequency dependence gives a model for the frequency response of the EO SOH modulator. It is compared to the measurement of the EO frequency response of a 1.5 mm long device presented in section 4.2. The next figure shows this comparison. The measurement deviates from the predicted behavior for frequencies larger than 10 GHz.

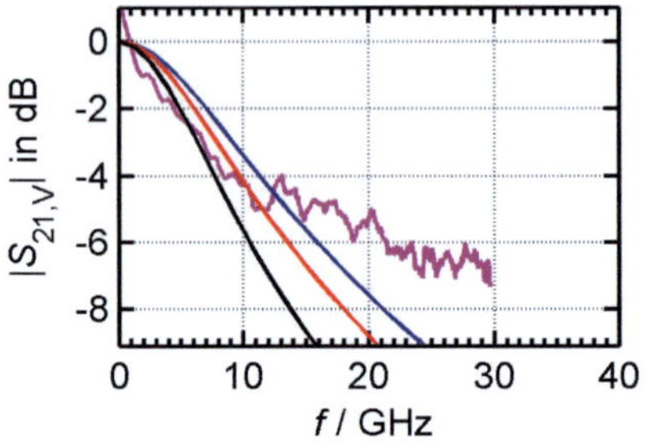

Fig. 2.20: Simulated (blue, red, black curves) and measured (magenta) frequency response of a 1.5 mm long EO SOH modulator. Frequency dependence of the attenuation of the RF wave according to the transmission line based model of [57], assuming an RC-limited BW of 10 GHz. Frequency dependence of the related effective length and the resulting absolute value of the EO S_{21} parameter. The blue, red and black curves correspond to an attenuation of (1, 3, 7) dB/mm at 40 GHz, respectively.

The deviation from the proposed model might be explained with a mix of various contributions of: (a) The substrate which has a low, but significant conductivity will cause attenuation because of induced currents and could be substituted with a high resistivity wafer and/or thicker buried oxide. (b) As described in [57] the RF wave propagates mostly in the CPW. However, there are also longitudinal currents propagating in the doped Si slabs. (c) To be able to use CMOS fabrication techniques and a complex stack of layers, there are many unavoidable planarization steps (chemical mechanical polishing) during fabrication to keep the top surface of the wafer planar. For reproducibility of the polishing process independent of the mask layout, the average density of structures in some layers needs to be the same for every mask. That is why a technique called tiling is employed. Wherever there is an unused area, e.g. in the Cu layer, thousands of little metal tiles with a certain spacing are placed, which will interact weakly with the propagating wave. Nevertheless, these tiles are going to introduce additional loss for higher frequencies. (d) Cross talk is avoided by using sufficient spacing between modulators.

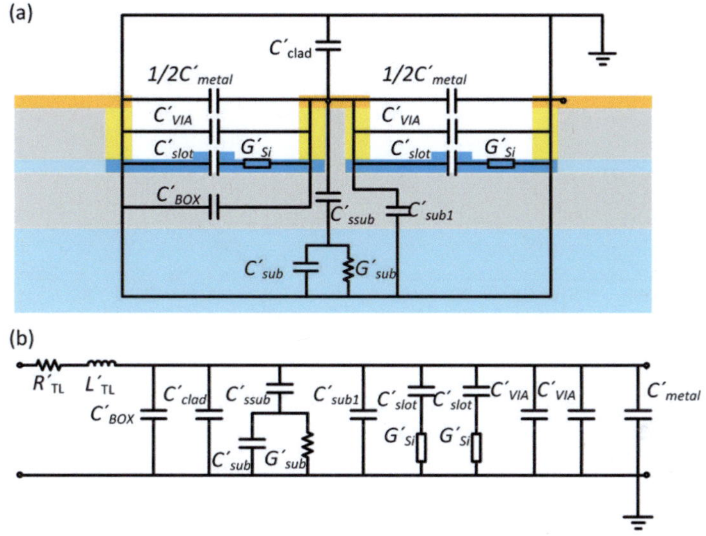

Fig. 2.21: Extended equivalent circuit model adapted from Yu et al. [58]. The following additional components are added: Capacitances related to the tungsten vias (GSG), the cladding (GSG), the buried oxide (BOX)(GSG), the substrate (S to substrate), taking a limited, additional conductance of the substrate into account. (a) Equivalent circuit components corresponding to the cross section are shown. (b) The RF wave is modeled with a series of infinitesimal long lumped-element circuits, as shown here.

Design conclusion

The limited bandwidth of the devices is mainly due to the RC limit. To further improve the BW of the devices, either the Si slabs have to be higher doped or a gate voltage has to be applied to create a better conductive electron accumulation layer.

The equivalent circuit model proposed by Witzens et al. can be extended according to Yu et al. [58], taking further elements into account, as shown in Fig. 2.21.

The influence of (a) the length l of the modulator, (b) the Si slab sheet conductivity[16] $\sigma_{n,Si}$ and (c) slot width w_{slot} on the RF power attenuation coefficient can be studied and discussed in detail with this equivalent circuit.

16 The sheet conductivity is given in units of kΩ/sq., where sq. means square, but is dimensionless and equal to 1.

The power attenuation coefficient is plotted with respect to $\sigma_{n,Si}$ and w_{slot} in Fig. 2.22.

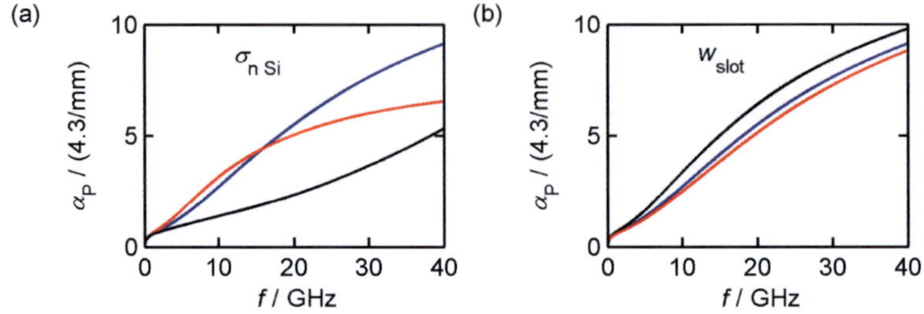

Fig. 2.22: Influence of Si slab sheet conductivity and slot width on the attenuation α_P of the RF wave, as simulated with the extended equivalent circuit model adapted from Yu et al. [58]. (a) The sheet conductivity of the Si slabs is varied (blue curve corresponds to measured resistance, i.e. 2.5 kΩ/sq., red curve 1.25 kΩ/sq., blue curve 20 kΩ/sq.). (b) The slot width w_{slot} is varied (blue curve 140 nm, red curve 100 nm, black curve 160 nm).

The resulting frequency response for the modulator of these parameter variations is shown in the next figure. Increasing the conductivity of the Si slabs well beyond the measured value appears to increase the model correlation with the measurements. The model for the frequency response obviously deviates from the measured frequency response, if all known measured values are inserted. It cannot be used to predict the absolute frequency response, but served well to study the influence of geometric and material parameters.

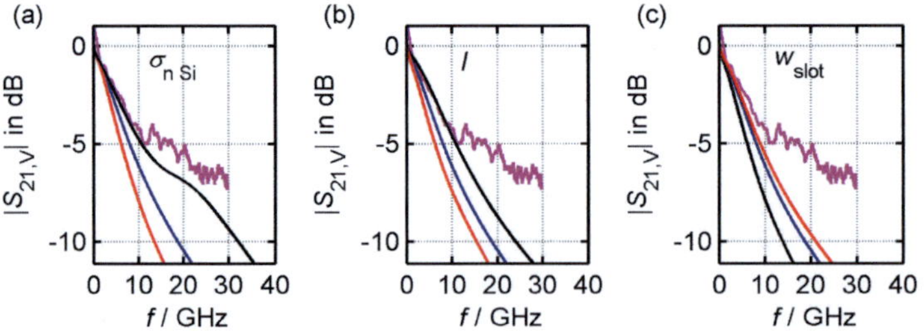

Fig. 2.23: Influence of selected parameters on the frequency response of an SOH MZM of 1.5 mm length, as simulated with the extended equivalent circuit model adapted from Yu et al. [58]. The measured response is plotted in magenta. (a) The sheet conductivity of the Si slabs is varied (blue curve corresponds to measured resistance, i.e. 2.5 kΩ/sq., red curve 1.25 kΩ/sq., blue curve 20 kΩ/sq.). (b) The length l is varied for reference (blue curve 1.5 mm, red curve 2 mm, black curve 1 mm). (c) The slot width w_{slot} is varied (blue curve 140 nm, red curve 100 nm, black curve 160 nm).

2.6 Modulation and Multiplexing Techniques

For digital data transmission over optical links a number of steps are necessary to electronically encode the data into symbols. These can be converted with an EO modulator to the optical domain for transmission over some distance before the optical signal is finally converted back to the electronic domain with a photodetector, to be sampled and decoded. Often, some pulse carving and/or filtering can be done in both the optical and the electrical domain. In this work, only the electrical-to-optical conversion is discussed (the EO modulator in the transmitter), and to a lesser extend the optical-to-electrical conversion (the optical part of the receiver).

Depending on the EO modulator (only called modulator from hereon), different properties of an incoming, optical carrier, here assumed to be a continuous wave (cw) carrier such as for example polarization, frequency, spatial mode and its complex amplitude can be controlled over time. The essential condition is that after modulation with any specific symbol, the light is in a state which can be distinguished from the states of all other symbols. It is a matter of choice how a bit stream is mapped to symbols and *modulated* onto an optical carrier and how an aggregation of bit streams is bundled to be *multiplexed*. For instance the polarization of a linearly polarized cw light can be modulated, or another property of two cw sources is modulated independently first, before a polarization multiplexing is done. Here, we select to modulate the complex amplitude of the light and do not consider multiplexing in frequency.

The decision about which modulation or multiplexing scheme is used is often driven by considerations about energy consumption and spectral efficiency.

2.6.1 Complex Amplitude Modulation

For a carrier of frequency f_c , expressed with the complex amplitude A or real valued amplitudes I & Q, the normalized electric field $E_{norm}(t)$ (absolute value normalized to be 1 when I,Q are 1, or -1) can be written as

$$E_{norm}(t) = A\exp(j2\pi f_c t) = I\cos(2\pi f_c t) + Qj\sin(2\pi f_c t) \ . \qquad (2.101)$$

A plot of the real-valued amplitudes Q vs. I is called a constellation diagram.

- A pure binary amplitude modulation between an on and an off state is called on-off-keying (OOK).
- A modulation of the complex amplitude with $I,Q \in \{|m_s|>0\}$ with $m_s \in \mathbb{N}$ and $m = m_s^2$ is called m-ary quadrature modulation (mQAM), i.e., there are m different symbols.
- A special case of mQAM is the modulation of the complex amplitude with $I,Q \in \{-1,1\}$, which is called quadrature phase shift keying (QPSK), because all symbols / states of light / constellation points could also be reached by just manipulating the phase of a cw carrier.

2.6.2 Complex Amplitude Demodulation

Depending on the modulation format, different detection schemes can be employed.

Direct detection

In this scheme, a photodiode serves to translate the incident power P_e of light into an electrical current i_{PD} which is proportional to the optical field intensity, and the responsivity (see Eq. (2.59)) with quantum efficiency η_{QE}, elementary charge e, energy of a photon $\hbar\omega$ or hf (h is the Planck constant, f the frequency of the light),

$$i_{PD} = \eta \frac{e}{hf} P_e \propto E^2 . \qquad (2.102)$$

When comparing the frequency response of MZMs for direct detection, it is useful to employ the 3 dB bandwidth (BW). A variation in voltage $\Delta V_{RF}(f)$ applied to the modulator operated at quadrature point delivers a variation in optical output power ΔP_{out} (which is proportional to i_{PD}). The 3 dB BW is defined as the frequency at which only half the RF power (which is proportional to i_{PD}^2) at the photodiode is generated at this frequency compared to the RF power at the photodiode at a reference frequency near 0 Hz (or the lowest relevant frequency used for modulation). In [60] this is defined as the electro-optic responsivity $R_{\mathrm{MZM,dB}}(f)$ of an MZM as

$$R_{\mathrm{MZM,dB}}(f) = 20\log_{10}\left(\frac{\Delta P_{\mathrm{out}}}{1\,\mathrm{W}}\frac{1\,\mathrm{V}}{\Delta V_{\mathrm{RF}}(f)}\right). \tag{2.103}$$

Coherent reception

To obtain the complete information of the complex amplitude of a signal, it can be compared to a local reference signal, a so called local oscillator (LO). This is called homodyne detection if the LO is locked in phase and frequency to the carrier, intradyne detection if the LO is not locked but still within the signal bandwidth, and heterodyne reception if the LO is not anymore within the signal bandwidth.

The reception can be done (homodyne, intradyne detection) by splitting the signal E_{S} and also the LO E_{LO} superposing both in two different settings (with a phase difference of 90°) using MMIs as shown in the next figure.

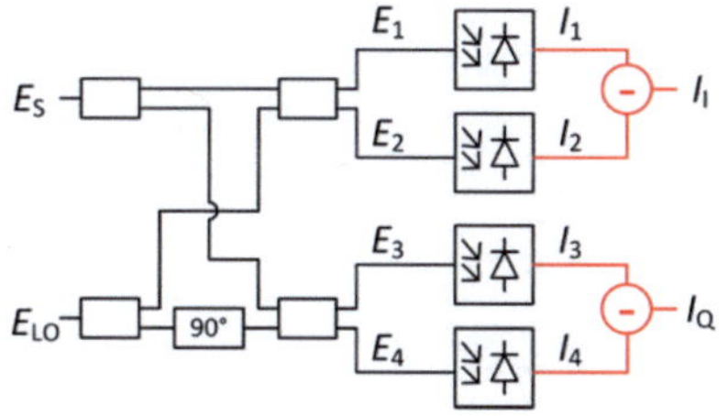

Fig. 2.24: Coherent receiver implemented using MMIs (depicted as rectangles, which split and combine optical WGs). Black lines denote optical WGs, red marks electric lines.

After the MMIs the electric fields can be written as

$$\begin{aligned} E_1(t) &= \left(E_\mathrm{S}(t) + E_\mathrm{LO}(t)\right)/\sqrt{2} \\ E_2(t) &= \left(E_\mathrm{S}(t) - E_\mathrm{LO}(t)\right)/\sqrt{2} \\ E_3(t) &= \left(E_\mathrm{S}(t) - jE_\mathrm{LO}(t)\right)/\sqrt{2}, \\ E_4(t) &= \left(E_\mathrm{S}(t) + jE_\mathrm{LO}(t)\right)/\sqrt{2} \end{aligned} \tag{2.104}$$

which eventually results in electric currents proportional to the intensities

$$\begin{aligned} I_\mathrm{I}(t) &= I_1(t) - I_2(t) = 2\Re\left\{E_\mathrm{S}(t)E_\mathrm{LO}^*(t)\right\}, \\ I_\mathrm{Q}(t) &= I_3(t) - I_4(t) = 2\Im\left\{E_\mathrm{S}(t)E_\mathrm{LO}^*(t)\right\}. \end{aligned} \tag{2.105}$$

Hence this detection scheme can be used to make a very linear transmission system employing MZMs or IQ modulators for advanced modulation formats, because of the linearity of the RF-to-optical up-conversion (MZM operated at minimum transmission point) and because of the linearity of the optical-to-RF down-conversion (received RF power is proportional to the voltage applied at the MZM).

As a consequence, when comparing the frequency response of modulators for coherent reception, it is useful to employ the 6 dB BW (using Eq. (2.103)), defined as the frequency at which a voltage applied to the modulator only delivers half the RF power at the output of the balanced photodiodes.

Despite the extra hardware requirements necessary for coherent detection, advanced modulation formats such as 16QAM are gaining interest, because of their spectral efficiency and because they can be used to realize transmission of higher bit rates at reasonable symbol rates (reasonable with respect to the driving electronics).

3 Light Emission with SOH Devices

This chapter presents the first demonstration of an SOH laser at infrared wavelengths. This chapter was submitted to a scientific journal [J1] (© for reprints will be transferred to this journal) and is printed here in its draft version.

[Start of Paper]

Lasing in Silicon-Organic Hybrid Waveguides

Dietmar Korn[1,*], Matthias Lauermann[1,*], Patrick Appel[1], S. Koeber[1,2], Luca Alloatti[1], Robert Palmer[1], Wolfgang Freude[1,2], Juerg Leuthold[1,2,3], Christian Koos[1,2]

[1] Institute of Photonics and Quantum Electronics (IPQ), Karlsruhe Institute of Technology (KIT), 76131 Karlsruhe, Germany

[2] Institute of Microstructure Technology (IMT), Karlsruhe Institute of Technology (KIT), 76344 Eggenstein-Leopoldshafen, Germany

[3] Now with Laboratory for Electromagnetic Fields and Microwave Electronics (IFH), Swiss Federal Institute of Technology (ETH), Zürich, Switzerland

[*] These authors contributed equally to this work.

Silicon photonics enables large-scale photonic-electronic integration by leveraging highly developed fabrication processes from the microelectronics industry. However, while a rich portfolio of devices has already been demonstrated on the silicon platform, on-chip light sources still remain a key challenge since the indirect bandgap of the material inhibits efficient photon emission and thus impedes lasing. Here we demonstrate a novel class of infrared lasers that can be fabricated on the silicon-on-insulator (SOI) integration platform. The lasers are based on the silicon-organic hybrid (SOH) integration concept and combine nanophotonic SOI waveguides with dye-doped organic cladding materials that provide optical gain. We demonstrate pulsed room-temperature lasing with on-chip peak output powers of up to 1.1 W at a wavelength of 1310 nm. The SOH approach enables efficient mass-production of silicon photonic light sources emitting in the near infrared (NIR) and offers the possibility of tuning the emission wavelength over a wide range by proper choice of dye materials and resonator geometry.

3.1 Introduction

Silicon photonics allows the fabrication of nanophotonic devices with commercial CMOS facilities and is therefore a highly attractive platform for large-scale photonic integration [61]–[63]. However, while a wide variety of silicon-based optical and electro-optical devices has been demonstrated over the last years [10], efficient on-chip light sources still represent a challenge [43] due to the indirect bandgap of silicon. Previously reported all-silicon light sources rely on stimulated Raman scattering as a gain mechanism [15], [16], but these devices require either strong pump lasers in combination with reverse-biased p-i-n-junctions or provide only limited output power in the microwatt range. Hybrid approaches, in which silicon is combined with direct-bandgap III-V compound semiconductors, allow for electrically pumped amplifiers [64] and lasers [65], [66], but fabrication requires sophisticated and technologically challenging die-to-wafer bonding processes or advanced technology for the direct growth of III-V quantum dots [67] on silicon. Regarding monolithic integration of light sources on silicon, an electrically pumped continuous-wave (cw) germanium-on-silicon laser has been demonstrated by using a combination of tensile strain and n-doping of the germanium to invoke direct-bandgap transitions in thin germanium layers that are grown on silicon substrates [18]. More recently, lasing has been shown without introducing mechanical strain by using a germanium-tin alloy [68] on silicon. However, fabrication of such devices still requires advanced crystal growth techniques. As an alternative, combinations of erbium-doped active cladding materials and SOI waveguides have been proposed [69], [70] and experimentally investigated [71], [72]. However, erbium features a rather small emission cross section and hence small gain. As a consequence, lasing in integrated erbium-clad devices has so far only been demonstrated for low-loss silicon nitride waveguides [73], but not for high index-contrast SOI waveguides. Regarding peak output power, even the most outstanding silicon-based lasers are currently limited to approximately a hundred milliwatts or less [67], [74], [75].

In this work, we demonstrate that lasing can be achieved by combining standard silicon-on-insulator (SOI) waveguides with dye-doped organic cladding materials. This concept of silicon-organic hybrid (SOH) integration is particularly well suited for flexible and low-cost mass-production of silicon photonic light sources emitting in the near infrared (NIR). In a proof-of-principle experiment, we demonstrate pulsed lasing at room temperature with peak output powers of up to 1.1 W at a wavelength of 1310 nm. Gain is provided by an NIR dye that was previously demonstrated to enable lasing in plastic waveguides [51]. More general, exploiting the virtually unlimited variety of organic optical cladding materials, SOH integration allows to complement silicon photonics with novel functionalities while still preserving the strengths of highly standardized CMOS processing [J6]. Our proof-of-principle demonstration of SOH light sources complements recent work on SOH integration, comprising high-speed all-optical signal processing [34], broadband electro-optic modulators [76], [77] and highly efficient low-power phase shifters [78].

3.2 Results

3.2.1 Concept and Fabrication of Silicon-Organic Hybrid Lasers

The basic idea of an SOH laser is illustrated in Fig. 3.1(a). The devices consist of SOI waveguides, which are terminated at both ends with Bragg reflectors [79] and which are covered by a fluorescent organic cladding material suitable for stimulated emission when optically pumped. For efficient light emission, the interaction of the guided optical mode with the active cladding must be maximized. This can be accomplished by using a narrow silicon strip waveguide, for which a large fraction of the guided mode reaches into the cladding, Fig. 3.1(b). Alternatively, a slot waveguide can be used, which consists of two closely spaced silicon rails, Fig. 3.1(c). In both cases, the dominant horizontal electric field component (E_x) of a quasi-TE mode experiences strong field discontinuities at the high-index-contrast sidewalls. For the slot waveguide, this leads to an especially pronounced field enhancement within the slot [80], and hence to a strong interaction with the active cladding.

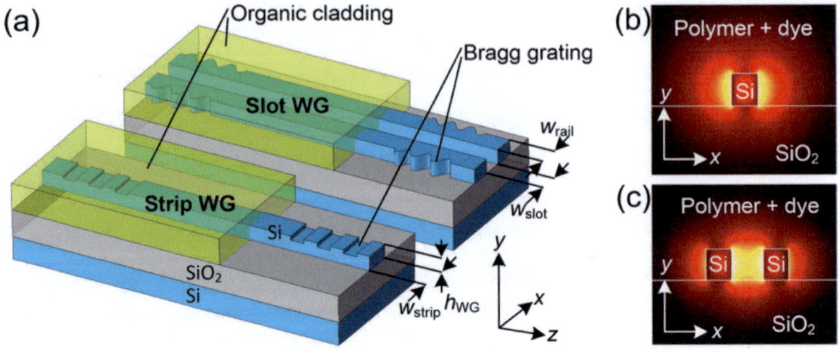

Fig. 3.1: Silicon-organic hybrid (SOH) laser: (a) Light is guided by silicon-on-insulator (SOI) strip or slot waveguides consisting of thin silicon nanowires (width w_{strip} = 150…500 nm, height h_{WG} = 200…350 nm) that are optically isolated from the silicon substrate by a thick oxide ($h_{\text{SiO2}} \approx 2$ µm). Optical gain is provided by a fluorescent organic cladding material, $h_{\text{clad}} \approx (500 \pm 50)$ nm, which entirely covers the strip or fills the slot ($w_{\text{rail}} \approx 100$ … 200 nm, $w_{\text{slot}} \approx 50$ … 200 nm). The optical pump is either launched from above or injected into the waveguide at one of the facets. Bragg reflectors can be used to provide wavelength-selective optical feedback. Interaction of the guided light with the active cladding is maximized by the design of the waveguides. (b) Dominant electric field component (E_x) of the fundamental quasi-TE mode for a narrow strip waveguide (color coding: lighter colors for higher magnitude). A large fraction of the guided mode propagates in the cladding. (c) Dominant electric field component (E_x) of the fundamental quasi-TE mode for a slot waveguide consisting of two tightly spaced silicon rails. Discontinuities of the dominant horizontal electric field component lead to a strong field enhancement within the slot region and hence to a strong interaction with the active cladding.

We prove the viability of the concept by investigating a simple test structure. To this end, strip and slot waveguides of 4.8 mm length were fabricated using a state-of-the-art SOI CMOS-based process [81]. The waveguides are embedded into a solid active cladding consisting of a poly(methyl methacrylate) (PMMA) matrix doped with 1 wt.% of the commercially available dye IR-26 [51], [52] having a maximum fluorescence at 1150 nm, see Supplementary Fig. 3.6. The cladding is deposited in a single post-processing step using standard spin-coating techniques. Scanning electron microscope (SEM) images of coated and uncoated samples can be found in Supplementary Fig. 3.7, showing that the PMMA cladding fills the slot completely without forming any voids. To enable laser operation in a wide wavelength range, we omit the wavelength-selective Bragg reflectors shown in

Fig. 3.1 and exploit spurious back reflection from cleaved waveguide facets and on-chip grating coupler (GC) structures, see Fig. 3.2(a) and (b). For the cleaved facets, power reflection factors of 4…8 % are estimated. Light emission from the cleaved facets is coupled to lensed standard single-mode fibers (SMF). For coarse alignment of the fibers, we use 1550 nm light coupled to the SOH waveguide via the GC. The GC is optimized for operation at a wavelength of 1550 nm and exhibit spurious back-reflection when operated at the laser emission wavelength of 1310 nm. This reflection amounts to a few percent and is comparable to that of the cleaved facet. A more detailed description of device fabrication can be found in Section 3.4.1.

Lasing could be demonstrated despite the comparatively low quality of the Fabry-Perot laser resonator, underlining the high potential of using dye-doped active claddings as gain media. In the experiment, the devices are pumped from above by a free-space line-focus beam using a pulsed laser with a wavelength of 1064 nm, a pulse duration of 0.9 ns (full width at half maximum, FWHM), and a pulse energy of up to 1.2 mJ at a repetition rate of 13.7 Hz. The experimental setup is explained in more detail in Section 3.4.2, which is followed by an estimation of the pump light absorbed in the active region of the lasers.

3.2.2 Characterization and Experimental Proof of Lasing

We measured the laser output power in the SMF as a function of the pump power for both the strip and the slot waveguide, Fig. 3.2(c) and (d). In both cases, a clear threshold can be observed at a launched average pump power of approximately 2.3 mW for the strip, and approximately 1.3 mW for the slot waveguide. The absorbed peak power at threshold in the vicinity of the waveguide can be roughly estimated to be 38 W for the strip, and 24 W for the slot waveguide, taking into account specific parameters of the individual waveguides, see Sections 3.4.2, 3.4.3 and 3.5.1 for more details.

The existence of the threshold indicates laser emission. The measured threshold level is in reasonable agreement with theory, see Section 3.4.5 for a more detailed discussion of the resonator properties and the associated laser emission characteristics. To rule out any laser look-alikes, we investigate further criteria formulated by Samuel et al. [82]. Below threshold, only amplified spontaneous emission (ASE) is to be seen, which increases exponentially with the pump power, see Insets of Fig. 3.2(c) and (d).

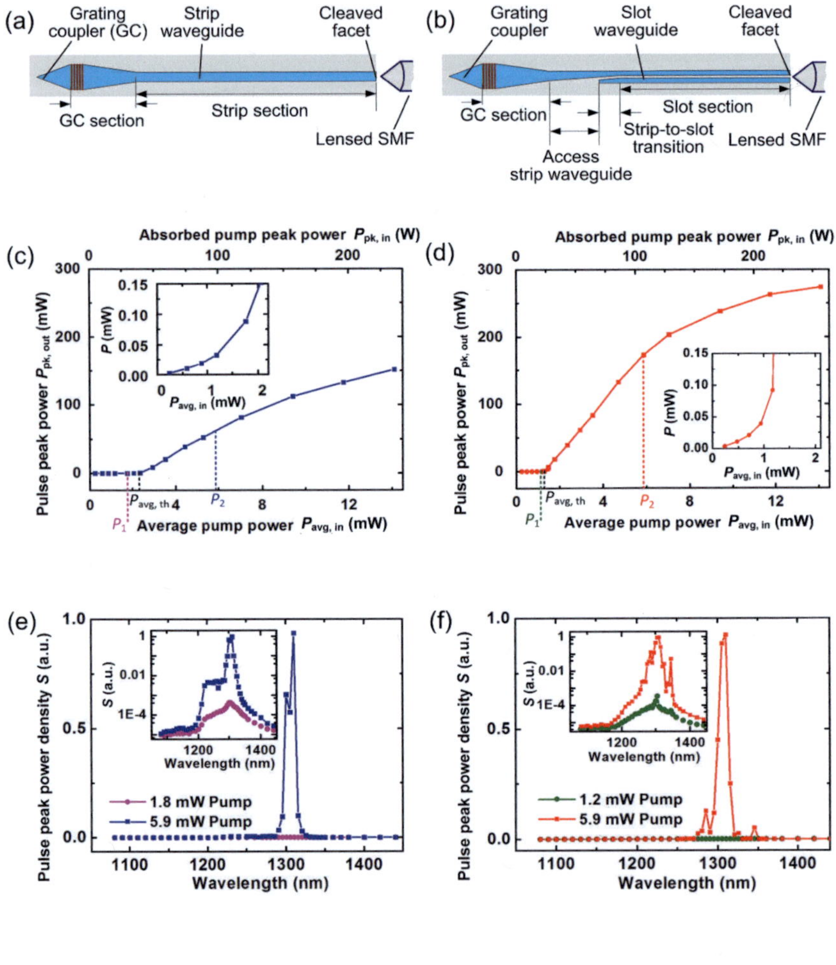
(a)
Grating coupler (GC)
Strip waveguide
Cleaved facet
GC section
Strip section
Lensed SMF
(b)
Grating coupler
Slot waveguide
Cleaved facet
GC section
Slot section
Strip-to-slot transition
Lensed SMF
Access strip waveguide
(c)
Absorbed pump peak power $P_{pk, in}$ (W)
Pulse peak power $P_{pk, out}$ (mW)
Average pump power $P_{avg, in}$ (mW)
$P_{avg, th}$
P_1
P_2
P (mW)
$P_{avg, in}$ (mW)
(d)
Absorbed pump peak power $P_{pk, in}$ (W)
Pulse peak power $P_{pk, out}$ (mW)
Average pump power $P_{avg, in}$ (mW)
$P_{avg, th}$
P_1
P_2
P (mW)
$P_{avg, in}$ (mW)
(e)
Pulse peak power density S (a.u.)
Wavelength (nm)
S (a.u.)
1.8 mW Pump
5.9 mW Pump
(f)
Pulse peak power density S (a.u.)
Wavelength (nm)
S (a.u.)
1.2 mW Pump
5.9 mW Pump

Fig. 3.2: Experimental proof of lasing in silicon-organic hybrid (SOH) strip and slot waveguides. The cladding consists of the commercially available dye IR26 [51] dispersed in a PMMA matrix. Cavity mirrors are formed by one cleaved waveguide facet and a grating coupler (GC). The GC is designed for coupling 1550 nm-light from an optical fiber to the strip and exhibits substantial back-reflection at the laser emission wavelength of 1310 nm. In both cases, the cavities are approximately 4.8 mm long. The laser output power is measured in a lensed single-mode fiber (lensed SMF) that collects light from the waveguide facet. (a) Strip waveguide consisting of a 450 µm long GC section and a 4.3 mm long strip section (waveguide height $h_{WG} \approx 220$ nm, width $w_{strip} \approx 210$ nm). (b) Slot waveguide comprising a 450 µm long GC section, a 235 µm long access strip waveguide, a 300 µm long strip-to-slot transition, and a 3.8 mm long slot waveguide section (rail width $w_{rail} \approx 180$ nm, slot width $w_{slot} \approx 215$ nm). (c) Peak output power $P_{pk,\,out}$ (all polarizations) in lensed SMF vs. illuminating average pump power $P_{avg,\,in}$ for the strip-waveguide cavity. A clear pump power threshold of $P_{avg,\,th} = 2.3$ mW can be observed. The measured incident average pump power (bottom scale) $P_{avg,\,in}$ is used to calculate the absorbed pulse peak power (top scale) $P_{pk,in}$, taking into account the specific parameters of the waveguide, see Sections 3.4.4, 3.5.1. The peak pump power was determined with a relative standard error of ± 14 %; for the peak power of the emitted pulse the relative standard error is ± 10 %, see Sections 3.4.3, 3.4.4. P_1 and P_2 denote the pump powers for which the spectra in Subfigure (e) are recorded. (d) Peak output power $P_{pk,\,out}$ in SMF vs. incident average pump power $P_{avg,\,in}$ for the slot waveguide cavity. A threshold pump power $P_{avg,\,th} = 1.3$ mW is found, which corresponds to 60 % of the threshold for the strip waveguide. P_1 and P_2 denote the pump powers for which the spectra in Subfigure (f) are recorded. (e) Emission spectra below (amplified spontaneous emission, ASE, magenta) and above threshold (blue) for the strip waveguide cavity. The spectrum is given in arbitrary units of the spectral pulse peak power density S recorded with a resolution bandwidth (RBW) of 5 nm (inset with logarithmic scale). Above threshold, the emission spectrum narrows considerably. (f) Emission spectra below (ASE, green) and above threshold (red) for the slot waveguide cavity (inset with logarithmic scale). The RBW amounts to 5 nm. Also here, the emission spectrum narrows considerably above threshold.

Above threshold, the output power increases linearly with the pump power. For very high pump powers, the laser power saturates. The saturation is attributed to absorption bleaching at the pump wavelength and to pump-induced free carrier absorption (FCA) in the SOI waveguide, see Section 3.4.6 for a more detailed discussion of optically induced losses.

Moreover, we investigate the emission spectra from the strip and slot waveguides below and above threshold, see Fig. 3.2(e) and (f). Broadband amplified spontaneous emission (ASE) can be observed for operation below threshold, see Insets of Fig. 3.2(e) and (f) (logarithmic scale). When pumped above threshold, the emission spectrum narrows considerably. In Fig. 3.2(e) and (f), the observed linewidth appears slightly larger than the resolution bandwidth of the spectrometer (RBW = 5 nm). We attribute this to a multitude of different longitudinal cavity modes which oscillate simultaneously at every pump pulse, see Section 3.5.2 and Supplementary Fig. 3.8 for a more detailed description.

Above threshold, the optical output of slot waveguides and of narrow strip waveguides is laterally single-mode, which can be inferred from the observation that there is a single well-defined optimum spot when coupling to a lensed SMF. For strip waveguides, lasing in higher-order lateral modes can be observed for waveguide widths of approximately 300 nm or more as reported in more detail in the next Section. For the devices shown in Fig. 3.2, the emitted light is predominantly polarized in the horizontal direction as is expected for lasing of the quasi-TE mode. The polarization extinction ratio (ER) is about 8 dB for both devices in Fig. 3.2. To confirm that the dye is indeed responsible for lasing, we prepared reference samples without dye in the PMMA cladding. These samples do not show noticeable light emission. Moreover, without the silicon waveguide but with dye in the cladding, only spontaneous emission is observed. These findings exclude any laser lookalikes and confirm the working principle of the SOH laser concept.

3.2.3 Influence of Waveguide Geometry

Regarding the influence of waveguide geometry on the performance of the SOH lasers, we find that lasing with high output powers can be achieved within a wide range of waveguide dimensions and that the output power is clearly related to the overlap of the guided mode with the active cladding. The geometry-dependent output power levels of different waveguide geometries are shown in Fig. 3.3. For the strip waveguide, we vary the width, Fig. 3.3(a), whereas for the slot waveguide, the rail width is fixed to 170 nm and the slot width is varied, Fig. 3.3(b). The length of the active section amounts to 4 mm for all devices. As before, the resonator is formed by back-reflection from a cleaved waveguide facet and from a grating coupler operated far from its design wavelength of 1550 nm. The experimental setup is the same as before and described in Section 3.4.2. The average pump power is fixed to 5 mW.

The colored areas of each bar represents the respective contribution of quasi-TE (blue) and quasi-TM polarization (green) to the total output power. For the strip waveguide, the laser power is largest when the strip width is smallest, i.e., when the light extends far into the active cladding. The second maximum at $w_{strip} = 375$ nm is due to lasing not only of the fundamental mode, but also of the next higher-order quasi-TE_{10} mode, which also strongly interacts with the cladding. The polarization extinction ratio (ER) reaches a maximum of (18±2) dB for the narrowest strip waveguides we investigated.

In Fig. 3.3(b), we consider slot waveguides and vary the slot width while keeping the rail width at a constant value of 170 nm. The TE mode dominates laser emission, since interaction with the cladding is enhanced by the electric-field discontinuities at the high-index-contrast sidewalls of the slot, as can be seen by comparing the field interaction factors [35] $\Gamma_{\text{clad, TE}}$ and $\Gamma_{\text{clad, TM}}$ for the two polarizations, see Table 3-1 and Section 3.5.1. Moreover, the emitted laser power increases with slot width. This is to be expected since larger slot widths lead to both larger field interaction factors of the guided mode within the active cladding and to a larger volume in which dye molecules can interact with the guided mode. For very large slot widths, the slot mode is only weakly guided and leaks into the high-index silicon substrate. As a consequence, the output power does not increase further. The polarization ER remains nearly constant and reaches a maximum of (8±2) dB for a slot width of 140 nm. For wider and narrower slots, the ER slightly is slightly smaller.

The optimum choice of the waveguide geometry depends on the desired balance between output power and polarization extinction ratio: High power output and a moderate ER when using slot waveguides have to be compared to about half the output power and a high polarization ER obtained from narrow strip waveguides. Using state-of-the-art CMOS fabrication, waveguide dimensions can be reproduced with tolerances of significantly less than 10 nm, which does not influence output power or polarization ER of the SOH lasers to a significant degree. SOH device performance can hence be expected to be resilient against fabrication inaccuracies.

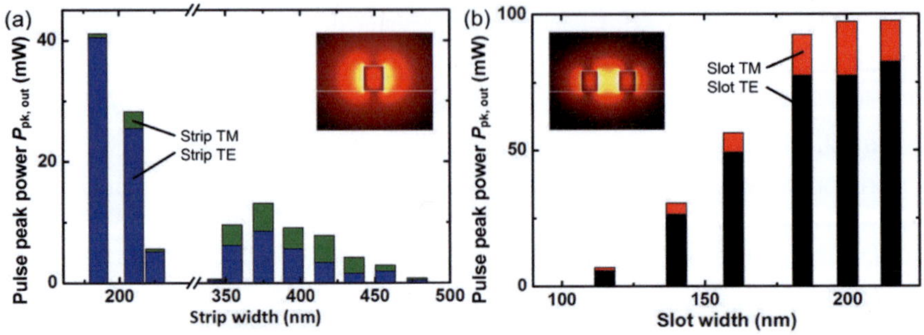

Fig. 3.3: Geometry-dependent peak output power $P_{\text{pk, out}}$ coupled into a lensed single-mode fiber for strip and slot waveguides. The resonator relies on back-reflection from one cleaved waveguide facet and from a grating coupler operated far from its design wavelength of 1550 nm. Quasi-TE and quasi-TM polarizations are measured separately. The average incident pump power is 5 mW for all samples. In the bar diagram, the differently colored areas represent the contributions of the quasi-TE and the quasi-TM polarization to the total output power; the total bar height corresponds to the total emission. Insets: Dominant electric field magnitudes of the fundamental quasi-TE modes. (a) Strip waveguide cavity. The laser power is largest when the strip width is smallest, i.e., when the guided light extends far into the cladding. The secondary maximum at 375 nm is due to lasing of the next higher-order mode (quasi-TE_{10}), which also has a strong overlap with the active cladding, but is not guided for smaller strip widths. (b) Slot waveguide cavity. An increase of the slot width leads to an increase of the field confinement in the cladding and to an expansion of the region in which the active dye interacts with the optical mode. As a consequence, the lasing power increases with slot width. For large slot widths, the fundamental mode is only weakly guided, and the laser power does not increase further. The rail width has less influence (not shown) and is fixed at 170 nm.

3.2.4 Dynamic Emission Behavior

The achievable peak output power of the SOH lasers is remarkable: For an SOH slot waveguide with cleaved facets on both sides, we measured peak output powers of up to 365 mW in the attached SMF, see Fig. 3.4(a), (b). The fiber-chip coupling losses are estimated to be (5 ± 1) dB, which leads to peak powers of (30.3 ± 1.0) dBm at the output facet, i.e., 1.1 W that could be coupled to an on-chip nanophotonic SOI waveguide. This is the highest peak power emitted from a silicon-based laser with on-chip cavity so far. A more detailed discussion of the estimation of emission power can be found in Section 3.4.3.

The time-dependent emission of the slot-waveguide laser is depicted in Fig. 3.4(c) for both polarizations, recorded at an average pump power of 5 mW. We observe laser emission into both the quasi-TE and quasi-TM mode,

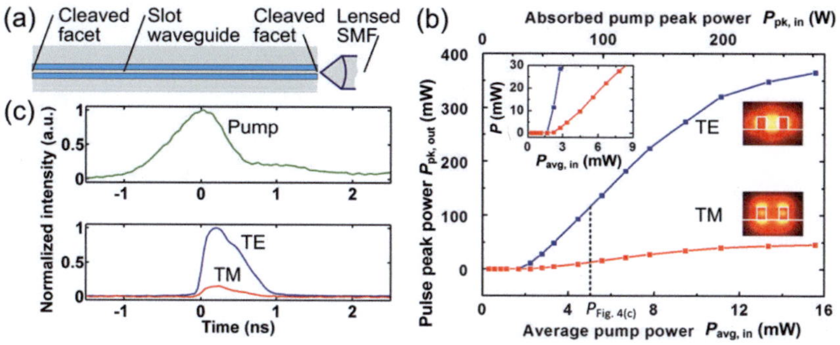

Fig. 3.4: Lasing in a silicon-organic hybrid (SOH) slot waveguide. In this experiment, cavity mirrors are formed by cleaved waveguide facets on both ends. The cavity length is 3.8 mm, the waveguide height amounts to 220 nm, and for the rail and the slot width, values of $w_{rail} = (160 \pm 15)$ nm and $w_{slot} = (180 \pm 15)$ nm were found. (a) Schematic top view of the slot waveguide. (b) Peak output power in the lensed SMF for quasi-TE and quasi-TM mode vs. incident average pump power. The absorbed pump peak power is estimated from the measured incident average pump power, see Section 3.4.4. Inset: Zoom-in of pulse peak power at low pump powers, demonstrating sharp thresholds for both TE and TM mode. (c) Temporal shape of the pump pulse at an average power of 5 mW (green) and of the corresponding emission pulses (TE, blue; TM, red). The shape of the pump pulse was measured by averaging over 16 pulses, and normalizing to a peak value of 1. Likewise, the emission pulses were measured in the SMF and averaged over 16 pulses. In the plot, the peak of the TE emission has been normalized to 1, and the TM emission is plotted at the same scale. The exact delay between pump and emission is unknown. The peak pump power was determined with a relative standard error of ± 14 %; for the peak power of the emitted pulse the relative standard error is ± 10 %, see Sections 3.4.3, 3.4.4.

which we attribute to local gain depletion: For large slot widths, the TE and TM modes occupy different cross-sectional domains of the active cladding, see insets in Fig. 3.4(b), and lasing may therefore occur simultaneously in both polarizations.

Since the overlap of the quasi-TE slot mode with the active cladding is larger than that of the TM mode, the TE mode experiences higher gain hence dominates lasing with a polarization extinction ratio of 9 dB. The TE and TM emission spectra are similar – see Supplementary Fig. 3.8 and Section 3.5.2 for a more detailed discussion.

Regarding the pulse shapes, we find that the mean FWHM duration of emission amounts to 0.6 ns, which is shorter than the pump pulse FWHM of 0.9 ns. Moreover, the emission pulse features an asymmetric shape and is delayed with respect to the pump pulse. The delay is attributed to the fact that laser emission can only set in once the pump intensity exceeds the threshold level. Note that the relative timing of pump pulse and emission pulses is subject to uncertainties of approximately ±100 ps due to different propagation delays in the fiber-based measurement setup, see Section 3.4.6 for more details. Therefore, the instantaneous pump power at the onset of laser emission cannot be directly associated with the threshold pump power level identified in Fig. 3.4(b). The asymmetric shape of the emission pulse might be caused by nonlinear absorption and subsequent relaxation processes in the active cladding. This aspect requires further investigation.

3.3 Discussion

SOH lasers have the potential to cover a broad range of different emission wavelengths between 1.1 μm to 1.6 μm by using suitable dye materials [83], [84]. Due to the high output power, the devices may even be used for exploiting nonlinear optic effects in nanophotonic waveguides. The SOH lasers are remarkably robust: We tested them repeatedly over several weeks without observing significant performance degradation in our experiments. This is in good agreement with previous observations, which have shown that photo bleaching of this dye can be neglected at our pump wavelength [51].

The devices presented in this paper are first-generation prototypes with considerable room for improvement. In particular, lasing threshold and linewidth of optical emission can be reduced by using optimized Bragg reflectors or ring resonators for optical feedback. Moreover, according to our study of the laser dynamics, we expect that better efficiency and lower threshold can be achieved by avoiding free-carrier absorption as an important loss mechanism of the cavity. To this end, one might consider dyes that allow for pumping wavelengths above the absorption edge of silicon [84]. Moreover, the pump efficiency can be improved considerably by guiding the pump light along the SOI waveguide to concentrate it in the active zone. This could be achieved by using an additional polymer waveguide around the SOI waveguide. High duty cycles or continuous-wave emission are in general difficult to achieve in dye lasers due to triplet-state excitation and subsequent photo-induced degeneration. This deficiency could be overcome by doping the matrix material with triplet-state quenching or triplet-trapping species of molecules [85], by using optofluidic concepts [86] or by choosing other gain materials such as lanlanthanide ions or colloidal quantum dots [87], [88] that might even be suited for direct electrical pumping [89].

Nevertheless, even without continuous-wave operation, SOH lasers enable greatly simplified one-step fabrication processes for realizing thousands of light sources directly integrated into silicon photonic circuitry. Such light sources lend themselves to, e.g., a wide range of applications in biosensing [90], where pulsed operation with low duty cycles is sufficient, where cost-efficient mass fabrication is essential to enable disposable chips for one-time use, and where pump efficiency is secondary. Moreover, the high peak power of the SOH lasers might open interesting opportunities in nonlinear infrared spectroscopy. Further investigation of the dynamics, optimization of the active cladding, and the use of better resonators should help enlarging the application range.

We therefore believe that the present approach will be the basis for a novel class of silicon photonic on-chip sources that stand out due to their high peak output power and ease of fabrication.

3.4 Methods

3.4.1 Fabrication of Silicon-Organic Hybrid Lasers

Waveguides were fabricated on silicon-on-insulator (SOI) wafers from SOITEC using a CMOS pilot line based on 193 nm deep-ultra-violet lithography [81]. All waveguides have a height of $h_{\mathrm{WG}} = 220$ nm and are optically isolated from the silicon substrate by a buried oxide (SiO_2) layer of thickness $h_{\mathrm{SiO2}} = 2$ µm.

The gain medium is deposited on the silicon waveguides in a single post-processing step by spin-coating. The active organic cladding consists of a poly(methyl methacrylate) (PMMA) matrix which is doped with 1 weight percent (wt.%) of the commercially available dye IR-26 [51]. The final thickness of the cladding amounts to $h_{clad} \approx (500 \pm 50)$ nm. The measured absorption and fluorescence spectra of a liquid dye solution are depicted [52] in Supplementary Fig. 3.6, exhibiting a fluorescence emission peak at 1130 nm. When dispersed in a solid, extended waveguide structure, the emission peak of IR26 shifts to approximately 1300 nm due to self-absorption along the waveguide in the overlap region of the emission and the absorption spectra [91].

This is in good agreement with the laser emission wavelength observed in Ref. [51].

3.4.2 Experimental Demonstration of Laser Emission

The experimental setup is depicted in Fig. 3.5. The SOH devices are pumped from top by a pulsed laser at a wavelength of 1064 nm with a duty cycle of approximately $p_t = 1.23 \times 10^{-8}$. The full width at half the maximum (FWHM) of the pump pulse amounts to 0.9 ns, the repetition frequency is 13.7 Hz. The incident pump power is controlled by adjusting the angle of a half-wave plate in front of a polarizing beam splitter (PBS). The pump light is polarized in a direction perpendicular to the waveguide axis and focused on the waveguide under test using a cylindrical lens (CL), see Fig. 3.5(a).

To measure emission from the SOH device, a lensed single-mode fiber (SMF) is placed near the facet, denoted as 'Fiber 2' in Fig. 3.5(a). The fiber collects the emitted light with an estimated coupling loss of approximately $(5 \pm 1)\,\mathrm{dB}$. By coupling an auxiliary light beam at 1550 nm through the on-chip grating coupler to the SOH waveguide using a second fiber (Fiber 1), we can facilitate the alignment of the lensed Fiber 2 with respect to the waveguide facet. Polarization-maintaining fibers are used throughout the setup. Fiber 2 is aligned such that the quasi-TE and quasi-TM emission of the SOH laser is coupled to the slow and the fast axis of the PM fiber, respectively.

To characterize the laser emission, we use two different detection paths in our setup: A "high-sensitivity detection" path, corresponding to the upper part in Fig. 3.5(b), and a "fast detection" path, represented by the lower part in Fig. 3.5(b).

The high-sensitivity path allows to measure input-output power characteristics and spectral properties of the laser emission. To this end, we use a monochromator and a highly sensitive photodetector with a large dynamic range, followed by an electrical low-pass filter for noise reduction and a standard oscilloscope, see Fig. 3.5(b). The oscilloscope is triggered by the emission of the pump laser and averages over 16 subsequent pulses. Due to the electrical low-pass filter and the bandwidth limitations of both the photodetector and the oscilloscope, the recorded electrical pulse is strongly widened compared to its optical counterpart. However, the peak of the recorded electrical pulse still remains proportional to the received optical power. This setup allows measuring the wavelength-resolved emission spectrum. For high output powers, an attenuator (not shown) was inserted in front of the photodiode.

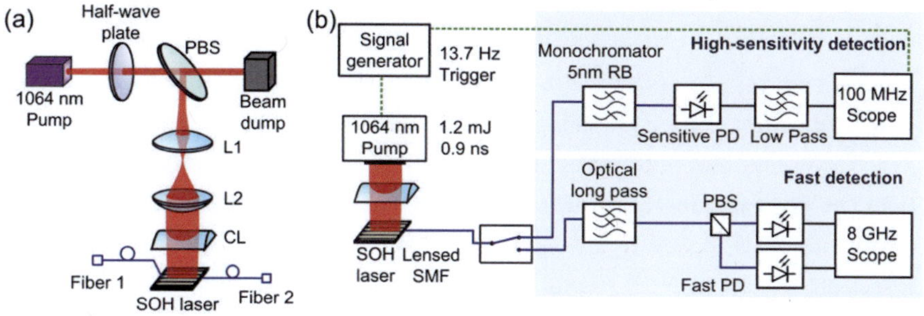

Fig. 3.5: Measurement setup. (a) Pump light at 1064 nm is focused on the SOH waveguide using a cylindrical lens. Pump power is adjusted by sending the linearly polarized light from the pump laser through a half-wave plate and a polarizing beam splitter (PBS). Fiber 1 (cleaved SMF illuminating a grating coupler) is used only to facilitate coarse alignment of fiber 2 (lensed SMF) by using 1550 nm light. (b) Emission from the SOH laser is collected by the lensed fiber (fiber 2), which is connected to different detector setups by an optical switch. The upper path is used for high-sensitivity detection. It contains a monochromator and a slow but highly sensitive photodetector (PD) to record weak fluorescence. The sensitive PD has a low bandwidth, and a consecutive electrical low-pass filter is used to further suppress noise. The lower "fast-detection" path is used for time- and polarization-resolved measurements. It is equipped with fast PDs. Residual pump power is blocked by an optical long-pass filter.

Time-resolved measurements are made with the fast-detection path: An optical long-pass filter blocks spurious pump light that might be scattered into the lensed fiber. A polarization beam splitter (PBS) separates the two polarization states for individual detection. Light pulses with a duration in the (sub-)ns-range are detected with fast photodiodes (NewFocus 25 GHz model 1434, NewFocus 45 GHz model 1014). A high-speed oscilloscope (Tektronix DPO 70804B, 8 GHz bandwidth, 25 GSa/s) is used to record time-resolved traces. The traces displayed in Fig. 3.4(c) have been obtained by averaging over 16 subsequent pulses. We find an average pump pulse duration of 0.9 ns FWHM with a standard error of 0.13 ns (15 %). Note that the cut-off wavelength $\lambda_{cut} = 1260$ nm of the fiber is above the wavelength of the pump light. Thus the pump pulse shape is likely influenced by modal dispersion.

The durations of the emitted SOH laser pulses are shorter than that of the pump pulse. For quasi-TE polarization, the mean FWHM duration amounts to 0.6 ns with a standard error of $\pm$ 0.06 ns (10 %).

3.4.3 Estimation of Emission Power Levels

For high output powers above the lasing threshold, the peak power levels in the output fiber were measured using the fast detection path of the setup depicted in Fig. 3.5, taking into account the responsivity of the fast photodiode and the optical and electrical losses of the various components. To obtain a lower boundary for the on-chip power levels, we assume that the total fiber-chip coupling losses are as low as 5 dB (factor 3.2). This value was estimated from reference measurements at 1550 nm; the actual losses at 1310 nm may be slightly higher. The coupling factor also includes losses of 6% due to reflection from the waveguide facet. A measured SOH laser peak power of 365 mW in the SMF hence corresponds to a power of at least 365 mW $\times$ 3.2 $\times$ 0.94 = 1.1 W which is coupled out from the waveguide facet and which could be used in an on-chip device that is connected to the SOH laser.

To estimate the variation of the measured emission power, the high-speed detection path depicted in Fig. 3.5(b) is used. We record subsequent emission pulses from a SOH slot waveguide similar to the waveguide depicted in Fig. 3.4(a), pumped well above threshold. A relative standard error of the pulse peak power of at most 10 % is measured.

For spectrally resolved measurements or for small power levels below the laser threshold, we use the high-sensitivity detection path of our setup. The peak power levels of the deformed pulses in the high-sensitivity path are calibrated by comparison to the corresponding peaks of the true pulse shapes in the fast detection path using medium power levels that can reliably be detected in both paths.

3.4.4 Estimation of Pump Power Levels

While the total average pump power is directly accessible by measurement, the absorbed peak pump power needs to be estimated based on further assumptions. The elliptical Gaussian pump spot features a major axis of 8 mm and a minor axis of 0.3 mm, both defined by the FWHM of the intensity on the chip surface. This is much larger than the active region of the SOH waveguide, i.e., the region in which pumped dye molecules interact with lasing waveguide mode. Considering the example of the device depicted in Fig. 3.4, the length $l_{\text{act, region}}$ = 3.8 mm of the active region is defined by the length of the slot waveguide section, and the width is estimated using the TE mode field diameter $\text{MFD}_x = 0.77$ µm in the lateral direction. The fraction of light that overlaps with the active zone is estimated by integrating the two-dimensional Gaussian distribution over the rectangle of MFD_x and waveguide length in the (x,z)-plane. This integral amounts to $p_{xz} = 0.0027$. To estimate the fraction p_y of pump light absorbed in the active cladding, we need to determine the corresponding absorption coefficient. From a direct transmission measurement using a 1.1 µm-thick IR26 dye-doped polymer layer on glass with the same dye concentration as the cladding material, the absorption cross section of the dispersed dye molecules is found to be $\sigma_p = 1.7 \times 10^{-16}\,\text{cm}^2$. This is in fair agreement with the value $\sigma_p = 5 \times 10^{-16}\,\text{cm}^2$ measured in a solution of the dye in 1,2-dichloroethane [92]. The thickness of the cladding h_{clad} = (500 ± 50) nm has been measured using a profilometer. Using $\sigma_p = 1.7 \times 10^{-16}\,\text{cm}^2$ and a dye molecule concentration of $N = 10^{19}\,\text{cm}^{-3}$, a value of $p_y = 1 - \exp\left(-\sigma_p N h_{\text{clad}}\right) = 0.08$ is found. The dye molecule density is derived from the measured mass ratio before mixing the PMMA matrix with the IR26 dyes. The total percentage of pump light absorbed in the active region is therefore $p_{xyz} = p_{xz} \times p_y = 0.022$ %. Using the measured pump pulse shape and the duty cycle, we find a ratio of average pump power to peak pump power of $p_{\text{avg/peak}} = 1.23 \times 10^{-8}$, which leads to a ratio of average incident pump power to *absorbed* peak pump power of $p = p_{\text{avg/peak}} / p_{xyz} = 5.6 \times 10^{-5}$. This ratio is used to relate the top and the bottom power scales in Fig. 3.4 (b). Consequently, the average incident threshold pump power of 1.8 mW leads to an estimate of the absorbed peak pump power of 32 W. The same method was used to relate the top and bottom power scales in Fig. 3.2(c) and (d); the corresponding ratios of average incident power to absorbed peak power are listed in Table 3-1. To estimate the variation of the measured pump power, a fraction of the pump pulse is coupled to a fiber and fed to a high-speed photo diode. From the measurements we find that the standard error of the pump is at most 14 %.

3.4.5 Resonator Characteristics and Threshold Pump Power

The measured threshold pump powers of the SOH lasers are in reasonable agreement with the losses of the cavities. This is demonstrated by analyzing the round-trip losses of a Fabry-Perot resonator with two cleaved facets as used in Fig. 3.4, and by relating them to the material gain of the active cladding.

The resonator round-trip losses are estimated by measuring the Fabry-Perot fringes in the transmission spectrum of the resonator and by evaluating the fringe contrast, see Section 3.5.3 for a more detailed discussion. For TE polarization, we find a contrast ratio C of approximately 0.5 dB between the transmission maxima and the adjacent minima, see Supplementary Fig. 3.9. According to Section 3.5.3, this corresponds to a total round-trip loss of 30.8 dB, calculated using $10\log_{10}(a^2R^2)$, where R denotes the power reflection factor at each facet and where a is the single-pass power transmission factor in the 3.8 mm-long waveguide. This result is in good agreement with a bottom-up consideration: We use a finite-element solver [93] to calculate the back-reflection R from the cleaved facet of an SOH waveguide, leading to a value of 6 % (−12.2 dB), see Table 3-1. Given the resonator length of $l = 3.8$ mm and the total round-trip loss of 30.8 dB, we hence estimate a propagation loss of approximately 0.9 dB/mm for the slot waveguide. This is in accordance with typically measured propagation losses of slot waveguides [94] which are of the order of 1 dB/mm.

At threshold, the round-trip losses of the resonator must be compensated by the round-trip amplification. For TE polarization, this requires a waveguide gain $\Gamma_{\mathrm{TE}}g = -\ln(aR)/l$ corresponding to 4.1 dB/mm, where $\Gamma_{\mathrm{clad,\,TE}} = 0.78$ denotes the field interaction factor of the guided mode with the active cladding, see Section 3.5.1 for more details. Laser emission in the dye cladding is governed by a transition that has a radiative lifetime [92] of the order of 14 ns and a fluorescence quantum efficiency ϕ ranging from 0.02% to 0.1%, see Refs. [92], [95]. The effective lifetime of the excited state hence amounts to $\phi\tau \approx 3\ldots14$ ps, thus much shorter than the durations of the pump and emission pulses. For estimating the pump intensity I_{thresh} at threshold, we may hence use steady-state approximations of the rate equations as described in Ref. [96] and Section 3.5.3.

This results in the relation

$$I_{\mathrm{thres}} = \frac{hc}{\lambda_p \sigma_{\mathrm{p}} \tau \phi} \left(\frac{\Gamma_{\mathrm{TE}} g}{\Gamma_{\mathrm{TE}} N \sigma_{\mathrm{e}} - \Gamma_{\mathrm{TE}} g} \right). \tag{3.1}$$

where $\lambda_p = 1064$ nm is the pump wavelength, $\sigma_p = 1.7 \times 10^{-16}\,\mathrm{cm}^2$ denotes the measured absorption cross section at this wavelength, N denotes the volume density of dye molecules, $\tau = 14.4$ ns is the excited-state lifetime, $\sigma_e = 0.5 \times 10^{-16}$ cm^2 is the emission cross-section [92], and ϕ is the fluorescence quantum efficiency with typical values ranging from 0.02% to 0.1%, see Refs. [92], [95], as specified for a liquid solution of the dye molecules.

When applied to the TE emission of the device depicted in Fig. 3.4, the previous equation leads to theoretically estimated threshold peak pump intensities ranging from 1.9 MW/cm^2 to 9.5 MW/cm^2. This is in reasonable with agreement our experimental estimation of the threshold peak pump intensity of 13.7 MW/cm^2. This estimation is based on the launched average threshold pump power of approximately 1.8 mW, the overlap $p_{xz} = 0.0027$ of the active area with the Gaussian pump spot in the (x,z)-plane, the pump duty cycle of approximately $p_t = 1.23 \times 10^{-8}$, and the area of the active zone having a length of $l = 3.8$ mm and a width of $\mathrm{MFD_x} = 0.77$ µm.

The deviation between the measured and the predicted the peak pump intensity is attributed to large uncertainties of the quantum efficiency ϕ. Previously published figures range from 0.02% to 0.1% and were measured in liquid dye solutions, see Refs. [92], [95], whereas we use the dyes in a solid polymer matrix. The measured 13.7 MW/cm^2 for the peak pump intensity can be reproduced when assuming a quantum efficiency of $\phi = 0.014$ %, which is comparable to the values obtained for liquid dye solutions. In addition, it turns out that free-carrier absorption may additionally increase the cavity losses, see Section 3.4.6. This would explain the fact that the experimentally measured threshold is slightly larger than the theoretically predicted value and lead to quantum efficiencies that are even closer to previously published values.

For TM polarization, the measured contrast of the Fabry-Perot fringes is comparable to that for TE polarization. Both polarizations hence experience similar cavity losses. Fig. 3.4(b) shows a slightly increased threshold pump power of the TM compared to the TE mode is attributed to a reduced field interaction factor of $\Gamma_{\mathrm{clad,\,TM}} = 0.42$ in the cladding compared to $\Gamma_{\mathrm{clad,TE}} = 0.78$. Moreover, the TM mode experiences higher FCA than the TE mode due to a stronger field interaction with the silicon waveguide core, see Section 3.4.6.

3.4.6 Optically Induced Losses and Dynamical Behavior

The dynamical behavior of the laser emission is depicted in Fig. 3.4(c). In this figure, the relative timing of the pump pulse and the emission pulses is subject to uncertainties: The various traces for the pump pulse, the TE emission and the TM emission were measured by an oscilloscope and a photodetector connected to the chip by standard single-mode fibers (G.652). For measuring the TE and the TM emission pulse, light was collected from the same fiber facet, and we may assume that both pulse trains experience the same propagation delay in the fiber. This is different for the pump - for measuring the pump pulse trace, we first had to remove the long-pass filter that was used to suppress residual pump light before it reaches the detector. We then moved the lensed fiber (Fiber 2 in Fig. 3.5) laterally to collect a small portion of 1064 nm pump light scattered from the surface of the chip. The group delay of the pump pulses from the fiber tip to the detector is slightly different than that of the emission pulses since the optical setup had to be changed slightly and since the optical fiber is operated below its single-mode cutoff wavelength of 1260 nm. This leads to higher-order mode propagation and hence to further uncertainties of the group delay. The overall uncertainty in relative timing between the pump and the emission pulses is estimated to be ±100 ps.

We also investigate the dynamics of intra-cavity losses at the emission wavelength of 1310 nm. The influence of two-photon absorption (TPA) of the emitted light and TPA-induced free-carrier absorption (FCA) can be neglected, see Section 3.5.4. As the only relevant loss mechanism, we identify FCA induced by direct absorption of 1064 nm pump light in the silicon waveguide core: During the pump pulse, free carriers accumulate within the core of the silicon waveguide, thereby leading to absorption and considerably increasing the optical losses of the resonator also at the emission wavelength.

For a rough quantitative estimate, we assume a linear absorption coefficient of $10\,\mathrm{cm}^{-1}$ for the 1064 nm pump light in the silicon waveguide core [32]. During pumping, photons absorbed in the waveguide create pairs of free carriers with an effective lifetime [94] in the order of 1 ns. Similarly to the considerations made for the active region of the SOH laser, the fraction of pump light that overlaps with the silicon waveguide is estimated to be $p_{\mathrm{xz,\,Si}} = 0.0011$, and the fraction of pump light absorbed in the 220 nm high silicon waveguide core is estimated to $p_{\mathrm{y,\,Si}} = 0.00022$. Using these values, the free-carrier density would reach $6.6 \times 10^{17}\,\mathrm{cm}^{-3}$ for an average pump power of 1.8 mW corresponding to the threshold of the laser depicted in Fig. 3.4. For this carrier density, an empirical model [32] allows us to roughly estimate an upper limit of the FCA-related propagation loss of approximately 5 dB/mm in the silicon core at the end of the pump pulse. Additional losses of this magnitude may significantly reduce the quality of the optical resonator during pumping and lead to an increased threshold. This is consistent with the observation that the experimentally measured threshold is slightly larger than the theoretically predicted value. We expect that in future devices, FCA can be mitigated by pumping at infrared wavelengths, which are not absorbed in the SOI waveguide core, or by using reverse-biased p-i-n structures that remove free carriers from the silicon core of the waveguides[15]. That would allow to considerably reduce threshold pump powers and to increase the slope efficiencies of the devices.

3.4.7 Summary of Resonator and Laser Characteristics

For the quantitative estimations in this paper, various waveguide and resonator parameters are used. These parameters are summarized in Table 3-1 along with threshold and emission power levels of the respective devices. The values are obtained either from experiments or from numerical simulations, e.g., for the case of the field confinement factor, effective area [97] and mode field diameter. The underlying mathematical relations are given in Section 3.5.1.

Property	Fig. 3.2(a) Strip waveguide	Fig. 3.2(b) Slot waveguide	Fig. 3.4 Slot waveguide
Height	$h_{\mathrm{WG}} = 220$ nm, $h_{\mathrm{clad}} \approx (500\pm50)$ nm		
Width	$w_{\mathrm{strip}} \approx 210$ nm	$w_{\mathrm{rail}} \approx 180$ nm $w_{\mathrm{slot}} \approx 215$ nm	$w_{\mathrm{rail}} = (160\pm15)$ nm $w_{\mathrm{slot}} = (180\pm15)$ nm
Length	$l_{\mathrm{complete}} = 4.8$ mm $l_{\mathrm{act.\ region}} = 4.3$ mm	$l_{\mathrm{complete}} = 4.8$ mm $l_{\mathrm{act.\ region}} = 3.8$ mm	$l_{\mathrm{complete}} = 3.8$ mm $l_{\mathrm{act.\ region}} = 3.8$ mm
Optical feedback	grating coupler + cleaved facet	grating coupler + cleaved facet	two cleaved facets
Effective refractive index (simulation)	$n_{\mathrm{eff}} = 1.71$	$n_{\mathrm{eff}} = 1.67$	$n_{\mathrm{eff}} = 1.61$
Facet power reflectivity for TE-mode (simulation)	8% -10.8 dB	6% -12.2 dB	6% -12.2 dB
Facet power reflectivity for TM-mode (simulation)	5% -12.6 dB	5% -13.0 dB	4% -13.6 dB
Propagation loss at 1310 nm	(12±5) dB/cm	(9±2) dB/cm	
Confinement factor (simulation)	$\Gamma_{\mathrm{clad,\ TE}} \approx 0.64$ $\Gamma_{\mathrm{clad,\ TM}} \approx 0.42$ $\Gamma_{\mathrm{Si,\ TE}} \approx 0.29$ $\Gamma_{\mathrm{Si,\ TM}} \approx 0.34$	$\Gamma_{\mathrm{clad,\ TE}} \approx 0.76$ $\Gamma_{\mathrm{clad,\ TM}} \approx 0.54$ $\Gamma_{\mathrm{Si,\ TE}} \approx 0.17$ $\Gamma_{\mathrm{Si,\ TM}} \approx 0.29$	$\Gamma_{\mathrm{clad,\ TE}} \approx 0.78$ $\Gamma_{\mathrm{clad,\ TM}} \approx 0.42$ $\Gamma_{\mathrm{Si,\ TE}} \approx 0.12$ $\Gamma_{\mathrm{Si,\ TM}} \approx 0.26$
Effective mode cross-section for third-order nonlinearities (simulation for TE-mode)	$A_{\mathrm{eff,\ clad}} = 0.15\ \mu\mathrm{m}^2$ $A_{\mathrm{eff,\ Si}} = 0.15\ \mu\mathrm{m}^2$	$A_{\mathrm{eff,\ clad}} = 0.20\ \mu\mathrm{m}^2$ $A_{\mathrm{eff,\ Si}} = 0.89\ \mu\mathrm{m}^2$	$A_{\mathrm{eff,\ cladd}} = 0.19\ \mu\mathrm{m}^2$ $A_{\mathrm{eff,\ Si}} = 1.56\ \mu\mathrm{m}^2$
Mode field diameter in x-direction (simulation for TE-mode)	$\mathrm{MFD}_x = 0.66\ \mu\mathrm{m}$	$\mathrm{MFD}_x = 0.79\ \mu\mathrm{m}$	$\mathrm{MFD}_x = 0.77\ \mu\mathrm{m}$
Ratio of average incident power to absorbed peak power	$p_{\mathrm{avg/peak}} / p_{\mathrm{xyz}} = 6.0 \times 10^{-5}$	$p_{\mathrm{avg/peak}} / p_{\mathrm{xyz}} = 5.5 \times 10^{-5}$	$p_{\mathrm{avg/peak}} / p_{\mathrm{xyz}} = 5.6 \times 10^{-5}$

Property	**Fig. 3.2(a)** **Strip waveguide**	**Fig. 3.2(b)** **Slot waveguide**	**Fig. 3.4** **Slot waveguide**
Launched average threshold pump power	2.3 mW	1.3 mW	1.8 mW
Absorbed peak threshold power	38 W	24 W	32 W
Laser emission peak power, at a wavelength of 1310 nm	150 mW (in SMF)	270 mW (in SMF)	365 mW (in SMF) 1.1 W (output facet) 1.2 W (in resonator)

Table 3-1: Waveguide properties, resonator parameters, and laser performance overview. All parameters refer to quasi-TE modes at a wavelength of 1310 nm, unless stated otherwise. The slot waveguides turn out to have a larger confinement of light to the cladding than the strip waveguides. This leads to a better interaction of the guided mode with the active cladding. The resonators in Fig. 3.2 include wide silicon waveguides sections. These sections consist of the grating couplers (GC) and access waveguides or transitions, in which the laser light is tightly confined to the Si waveguide core and hence only a very minor part interacts with the active cladding. Consequently, only the narrow strip section or the slot section contribute to lasing and are regarded as part of the active region. Therefore, the device lengths l_{complete} and the lengths $l_{\text{act. region}}$ of the active region are stated separately.

3.5 Supplementary Information

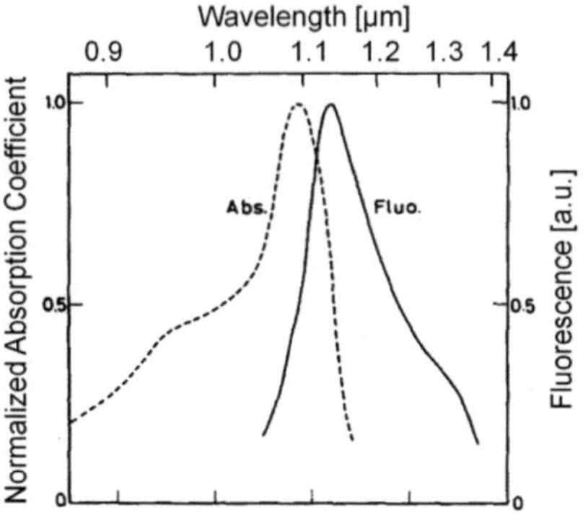

Supplementary Fig. 3.6: Absorption and fluorescence spectra of the organic dye IR-26 [52], dissolved in 1,2-dichloroethane. The exact shapes of these spectra depend on the host material. Due to self-absorption in the cladding of the silicon waveguides, the emission peak is shifted towards 1300 nm compared to the depicted fluorescence spectrum in solution. In our experiment, the material is pumped at a wavelength of 1.064 µm, thus close to the wavelength of maximum absorption of 1080 nm. Image reproduced from [52].

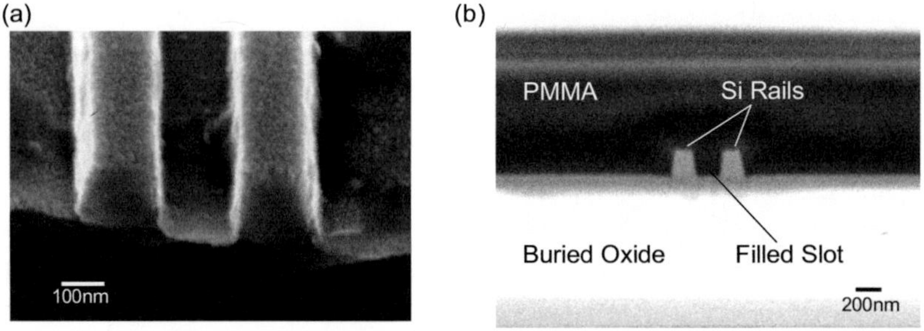

Supplementary Fig. 3.7: Scanning electron microscope (SEM) images of fabricated slot-waveguide samples. The structures are nominally identical to the one used for the SOH laser in Fig. 3.4. (a) Cleaved facet of an SOI slot waveguide after removing the PMMA cladding. (b) Cross-sectional view of a SOH slot waveguide coated by an undoped PMMA cladding. The figure was obtained by focused ion beam (FIB) milling and scanning electron microscopy. The cladding was deposited by spin coating and fills the slot completely without forming any voids.

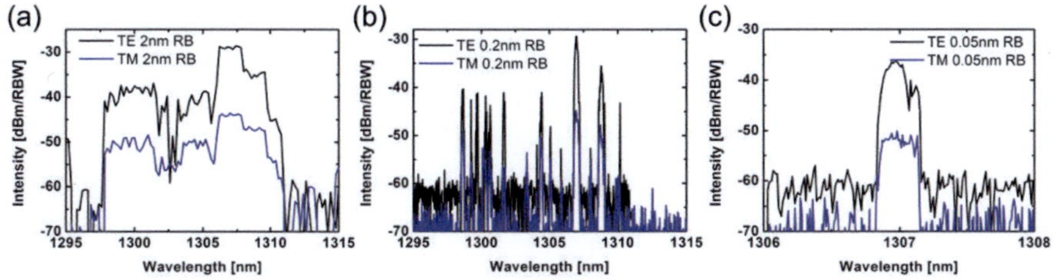

Supplementary Fig. 3.8: High-resolution laser emission spectra in TE and TM polarization derived from the slot-waveguide laser that was also used in Fig. 3.2. Fiber-chip coupling has been re-optimized for each polarization-resolved measurement. We use an optical spectrum analyzer operated at a hold time of 170 ms for each spectral point. As a consequence, at least two pulses at a repetition rate of 13.7 Hz are recorded at each wavelength. (a) Overview spectrum at a resolution bandwidth (RBW) of 2 nm. (b) Scan with 0.2 nm resolution revealing a multitude of underlying narrowband spectral lines, which we attribute to longitudinal cavity modes. Repeated scans show peaks at positions that are indistinguishable from scan to scan within the measurement accuracy. We conclude that the same set of longitudinal modes starts lasing simultaneously in each pulse and contributes to the overall output power. (c) Scan with 0.05 nm resolution. A single emission peak of one longitudinal mode exhibits a linewidth of around 0.2 nm. We attribute this to chirp-induced spectral broadening due intra-cavity free-carrier dynamics induced by absorption of 1064 nm pump light in the silicon waveguide cores.

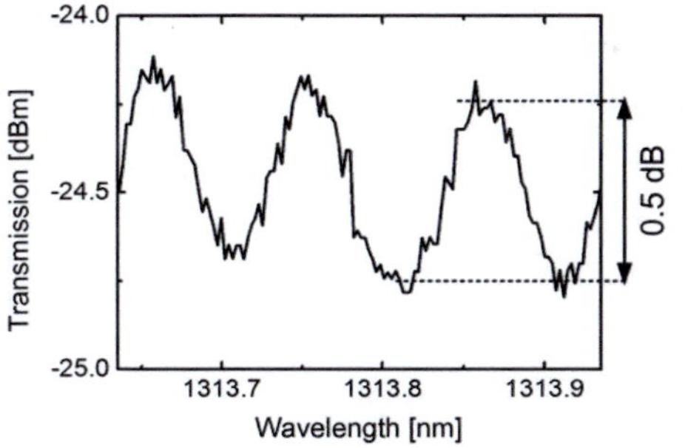

Supplementary Fig. 3.9: Close-up of a transmission spectrum of a slot waveguide with cleaved facets, measured at TE polarization. The free spectral range corresponds to a Fabry-Perot cavity of 3.75 mm with a group refractive index of 2.1. The contrast of the fringes of 0.5 dB corresponds to a roundtrip loss of 30.8 dB.

3.5.1 Waveguide Properties

The observed output power of the slot-waveguide laser, Fig. 3.2(d), is larger than the output power of the strip waveguide, Fig. 3.2(c), and the lasing threshold is also lower for the slot waveguide. To understand this behavior, not only the resonator but also the available gain has to be considered. The available gain and dynamic loss depend on the distribution of light in the waveguide cross-section. The overlap of the guided light with the active organic cladding can be quantified by means of the field interaction factor given by [98]

$$\Gamma = \frac{\int\limits_{\text{reg}} \frac{n_{\text{reg}}}{Z_0} \left|\tilde{\mathbf{E}}_{\text{t}}(x,y)\right|^2 \text{d}x\text{d}y}{\int\limits_{\text{all}} \Re\left\{\tilde{\mathbf{E}}_{\text{t}}(x,y) \times \tilde{\mathbf{H}}_{\text{t}}^{*}(x,y)\right\} \cdot \mathbf{e}_z \text{d}x\text{d}y}. \tag{3.2}$$

In this relation, the refractive index of the active polymer cladding (reg, integration region of the numerator integral) is denoted as n_{reg}, Z_0 is the vacuum wave impedance, $\mathbf{e}_z$ the unit vector in z-direction, and $\tilde{\mathbf{E}}_{\text{t}}(x,y)$ and $\tilde{\mathbf{H}}_{\text{t}}(x,y)$ denote the electric and magnetic mode fields in the transverse (x,y)-plane. The integration domain of the numerator integral corresponds to the active cladding of the waveguide, whereas the denominator extends over entire waveguide cross section. When calculating the field interaction to the active cladding region according Eq. (3.2), we find that the slot waveguide of Fig. 3.2(b) exhibits an interaction factor $\Gamma_{\text{clad, TE}} \approx 0.76$ that is slightly larger than the value obtained for the strip waveguide $\Gamma_{\text{clad, TE}} \approx 0.64$ in Fig. 3.2(a) for the cladding, see Table 3-1 in Section 3.4.7.

For nonlinear optical effects such as two-photon absorption, the concentration of light within a certain part of the waveguide, e.g., the silicon waveguide core, must be quantified. This is usually done by means of the effective area of third-order nonlinear interaction. The effective area for interaction with a certain cross-sectional region (reg) of the waveguide is given by [99]

$$A_{\text{eff}} = \frac{\left(\int\limits_{\text{all}} \Re\left\{\tilde{\mathbf{E}}_{\text{t}}(x,y) \times \tilde{\mathbf{H}}_{\text{t}}^{*}(x,y)\right\} \cdot \mathbf{e}_z \text{d}x\text{d}y\right)^2}{\int\limits_{\text{reg}} \frac{n_{\text{reg}}^2}{Z_0^2} \left|\tilde{\mathbf{E}}_{\text{t}}(x,y)\right|^4 \text{d}x\text{d}y}. \tag{3.3}$$

For the strip waveguide, we find that the effective area of third-order non-nonlinear interaction in the silicon core amounts to $A_{\text{eff, Si, strip}} = 0.15\ \mu\text{m}^2$ and is much smaller than its slot-waveguide counterpart, $A_{\text{eff, Si, slot}} = 0.89\ \mu\text{m}^2$. Hence, for the same power of guided light, any non-linear effects in the waveguide core such as two-photon absorption will be much stronger in the strip waveguide than in the slot waveguide as discussed in Section 3.5.4.

For describing the spatial extent of the gain region of an active SOH waveguide, we use the second-moment width $\text{MFD}_x = \text{D}4\sigma$, i.e., four times the second central moment along the x-direction,

$$\text{MFD}_x = 4\sqrt{\frac{\int_{\text{reg}} \frac{n_{\text{reg}}}{Z_0} \left|\tilde{\mathbf{E}}_\text{t}(x,y)\right|^2 (x-\bar{x})^2 \,\text{d}x\text{d}y}{\int_{\text{all}} \Re\left\{\tilde{\mathbf{E}}_\text{t}(x,y) \times \tilde{\mathbf{H}}_\text{t}^*(x,y)\right\} \cdot \mathbf{e}_z \text{d}x\text{d}y}}. \tag{3.4}$$

Note that the first moment is zero, because the waveguides and the magnitude of the resulting electric mode fields are symmetric with respect to the (y,z)-plane. The mode field diameter is $\text{MFD}_x = 0.66\ \mu\text{m}$ ($\text{MFD}_x = 0.79\ \mu\text{m}$) for the strip (slot) waveguide in Fig. 3.2, see Table 3-1. Hence we may expect the number of excited dye molecules available for stimulated emission to be slightly larger for the slot waveguide than for the strip waveguide.

3.5.2 High-Resolution Spectra

To explain the comparatively large bandwidth of laser emission in Fig. 3.2, we have investigated the emission spectrum at higher resolutions of 2 nm, 0.2 nm, and 0.05 nm, see Supplementary Fig. 3.8(a) to (c). We investigated both TE and TM polarization. The hold time of the optical spectrum analyzer at each measurement point amounts to 170 ms, and is chosen such that at least two emission pulses at a repetition rate of 13.7 Hz are recorded. We find that the spectra consist of a multitude of narrowband spectral lines, which we attribute to different longitudinal modes of the cavity that exhibit laser emission simultaneously in each cycle. The positions of these emission peaks are reproducible when measuring the spectrum repeatedly. A zoom-in of a single emission peak is depicted in Supplementary Fig. 3.8(c), exhibiting a FWHM linewidth of 0.1 … 0.2 nm. This linewidth is much larger than the 0.003 nm FWHM expected for an unchirped pulse of 1 ns duration. We attribute the excess spectral bandwidth to a strong chirp, induced by free-carrier dynamics of the cavity which originate from absorption of 1064 nm pump light in the silicon waveguide cores and which lead to a strong change of the cavity refractive index during pulse emission. The fact that both the TE and the TM emission exhibit spectral narrowing above threshold is another strong indication for laser emission into both modes. A striking similarity of the TE and TM spectra is observed at highest resolution, Supplementary Fig. 3.8(c). We consider this as an indication that the TE and TM modes are coupled: TE starts lasing at lower pump powers, and a small portion of the TE light will couple to the TM mode and act as a seed for lasing.

3.5.3 Resonator Loss and Pump Threshold

Estimation of resonator losses

For a Fabry-Perot resonator, the phase shift δ accumulated during one round trip can be calculated according to

$$\delta = 2n_e l \frac{2\pi}{\lambda}, \tag{3.5}$$

with n_e denoting the effective refractive index in the waveguide of length l and λ being the vacuum wavelength. The transmission through the resonator depends on the incident wavelength and the resonator parameters. Let R denote the facet power reflectivity and a the single-pass power transmission factor. The wavelength-dependent power transmission T of the Fabry-Perot resonator can then be written as

$$T = \frac{(1-aR)^2}{1+a^2R^2-2aR\cos\delta}. \tag{3.6}$$

From this relation, we can derive an expression for the fringe contrast C, i.e., the ratio of the power transmission maxima and the adjacent minima, similar as for the Hakki-Paoli method [100]

$$C = \frac{(1+aR)^2}{(1-aR)^2}. \tag{3.7}$$

The fringe contrast C is obtained from a high-resolution transmission spectrum of the resonator, see, e.g., see Supplementary Fig. 3.9. Solving Eq. (3.7) for the single-pass power transmission factor aR of the Fabry Perot resonator, we obtain -15.4 dB when evaluating 10 log(aR) for the slot waveguide presented in Fig. 3.4. This corresponds to a 30.8 dB round-trip loss. We can also determine the coefficient of finesse F of the resonator using the relation

$$F = \frac{4aR}{\left(1 - aR\right)^2} \tag{3.8}$$

For the slot waveguide presented in Fig. 3.4, the coefficient of finesse is 0.12.

Calculation of pump threshold

Assuming a quasi-four-level system and neglecting the triplet states, the simplified rate equation can be written as [96]

$$\frac{\mathrm{d}N_1}{\mathrm{d}t} = N_0\left(\sigma_\mathrm{p}\frac{I_\mathrm{p}}{hc/\lambda_\mathrm{p}} + \sigma_a\frac{I_\mathrm{e}}{hc/\lambda_\mathrm{e}}\right) - N_1\left(\sigma_\mathrm{e}\frac{I_\mathrm{e}}{hc/\lambda_\mathrm{e}} + \frac{1}{\tau\phi}\right) \tag{3.9}$$

In this equation, the volume density of dye molecules in the ground state (level 0) is denoted by N_0, whereas N_1 denotes the volume density of dye molecules in the excited level (level 1). The total density of dye molecules is $N = N_1 + N_0$. The wavelength and the intensity of the emitted light are given by λ_e and I_e, respectively, and λ_p and I_p denote the corresponding quantities for the pump light. Plank's constant is denoted as h and the velocity of light as c. The quantity τ is the decay time associated with radiative transitions from the excited state to the ground state, and ϕ is the fluorescence quantum yield. The emission cross section is denoted as σ_e, the absorption cross section at the emission frequency is called σ_a and the absorption at the pump frequency is σ_p.

For 1 wt% dye in polymer, the total volume density of dye molecules amounts to $N = 1 \times 10^{19}\,\mathrm{cm}^{-3}$. For the excited state transition lifetime and the emission cross section at 1310 nm we use values from literature [101], $\tau = 14.4$ ns and $\sigma_\mathrm{e} = 0.5 \times 10^{-16}\,\mathrm{cm}^2$. Reabsorption at the emission wavelength can be neglected, $\sigma_\mathrm{a} = 0$. The absorption cross section σ_p at the pump wavelength λ_p of 1064 nm was obtained from a transmission measurement of a dye-doped polymer film on a glass substrate and amounts to $\sigma_\mathrm{p} = 1.7\cdot10^{-16}\,\mathrm{cm}^2$.

For $\phi \approx 0.1\%$, the decay of excited states is governed by the time constant $\tau\phi \approx 14\,\mathrm{ps}$. This is much shorter than the duration of the 0.9 ns pump pulse, and we may hence consider the steady-state behavior of Eq. (3.9) to estimate the threshold intensity. Assuming further that the emission intensity I_e is still zero at threshold, the density of excited state molecules is obtained from Eq. (3.9),

$$N_1 = \frac{N\sigma_p I_p \tau\phi}{\sigma_p I_p \tau\phi + hc / \lambda_p}. \tag{3.10}$$

At this density, the single-pass gain must compensate the single-pass losses in the resonator, thus $\Gamma_{clad} N\sigma_e = \Gamma g = -\ln(aR)/l$, where Γ_{clad} denotes the field interaction factor as defined in Eq. (3.2). The threshold pump intensity is thus obtained to

$$I_{p,thres} = \frac{hc}{\lambda_p}\frac{1}{\tau\phi\sigma_p}\left(\frac{-\ln(aR)}{l\Gamma_{clad} N\sigma_e + \ln(aR)}\right). \tag{3.11}$$

For the slot waveguide depicted in Fig. 3.4, we experimentally determine a launched peak pump threshold intensity of $I_{p,thresh} = 13.7$ MW/cm². This estimation is based on the launched average threshold pump power of approximately 1.8 mW, the overlap $p_{xz} = 0.0027$ of the active area with the Gaussian pump spot in the (x,z)-plane, the pump duty cycle of approximately $p_t = 1.23 \times 10^{-8}$, and the area of the active zone having a length of $l = 3.8$ mm and a width of $MFD_x = 0.77$ µm. Using these parameters, the experimentally determined peak pump threshold intensity of $I_{p,thresh} = 13.7$ MW/cm² can be reproduced by Eq. (3.11) if we assume a quantum efficiency of 0.014 %. This value is close to published quantum efficiencies ranging from 0.02% to 0.1%. The remaining deviations are attributed to the fact that the references refer to a liquid solution of the dye molecules rather than to a solid polymer matrix as used in our experiments. Moreover, free-carrier absorption contributes additional cavity loss and leads to an increase of the pump threshold, see Section 3.4.6.

3.5.4 Optically Induced Nonlinear Losses

Losses for light propagating in silicon at the emission wavelength of $\lambda_e = 1300$ nm comprise linear waveguide losses, two-photon absorption (TPA), and absorption by free carriers that are generated as a result of TPA. The associated decay of intensity I along the propagation direction z can be approximated by a first-order differential equation [102],

$$\frac{dI(z)}{dz} = -\left(\alpha_{\text{lin}} + \beta_{\text{TPA}} I(z) + \alpha_{\text{FCA}} I^2(z)\right) I(z). \tag{3.12}$$

In this relation α_{lin} denotes the linear propagation loss in the SOI waveguide, $\beta_{\text{TPA}} = 0.74\,\text{cm/GW}$ denotes the TPA coefficient of bulk silicon [103], and α_{FCA} is the coefficient of TPA-induced FCA [102],

$$\alpha_{\text{FCA from TPA}} = 1.45\times 10^{-21}\,\text{m}^2 \left(\frac{\lambda_\text{e}}{1.55\,\mu\text{m}}\right)^2 \frac{\beta_{\text{TPA}}\tau_{\text{eff, Si}}\lambda_\text{e}}{2hc}, \tag{3.13}$$

where $\tau_{\text{eff, Si}}$ denotes the effective free-carrier lifetime in the silicon waveguide core and amounts to $\tau_{\text{eff, Si}} = 1\,\text{ns}$ [104]. Note that this lifetime is much shorter than the carrier lifetime in bulk silicon due to increased recombination rates at the etched waveguide surfaces.

For an exemplary estimation of the influence of two-photon absorption of 1310 nm emission, we consider the slot waveguide of Fig. 3.4 and assume an emitted laser peak power of $P_{\text{facet}} = 1.1\,\text{W}$ just outside the facet. The actual power inside the laser resonator is even higher due to the 6 % power reflectivity of the facet and is estimated to be approximately 1.2 W. Only a part of the light is guided in the silicon waveguide core with an effective mode cross section of $A_{\text{eff, Si}} = 1.56\,\mu\text{m}^2$, see Table 3-1. To estimate an upper boundary for the TPA losses, we calculate the intensity assuming that all power is distributed homogenously over this area. This leads to an intensity estimate of $I_{1310\,\text{nm}} = 0.077\,\text{GW/cm}^2$, resulting in a TPA-induced loss contribution of 0.03 dB/mm for this specific power level. Compared to linear losses in the waveguide of 0.9 dB/mm, TPA is thus negligible as a direct loss mechanism in our case. This leads to $\alpha_{\text{FCA from TPA}} = 2.5\times 10^{-23}\,\text{m}^3\text{W}^{-2}$, which is the TPA-induced FCA coefficient, and thus to a loss contribution of less than 0.06 dB/mm at peak of the emitted laser pulse. The contribution of TPA-induced FCA to waveguide losses can hence also be safely neglected. In contrast to that, FCA induced by direct absorption of the 1064 nm pump light turns out to be a relevant effect.

[End of Paper]

4 Modulation with SOH Devices

In this part of the thesis, we report on characterization results for SOH modulators using claddings from two different material classes. A scenario for combining SOH modulators with electronic driver circuits is discussed. The SOH platform is shown to be suited to implement modulators.

4.1 MZM Modulator Based on Organic Crystals

[Start of Paper, *IEEE Photonics J.*, vol. 6, no. 2, pp. 1–9, Apr. 2014]

Electro-Optic Organic Crystal Silicon High-Speed Modulator

D. Korn[1], M. Jazbinsek[2], R. Palmer[1], M. Baier[1], L. Alloatti[1], H. Yu[3], W. Bogaerts[3], G. Lepage[4], P. Verheyen[4], P. Absil[4], P. Guenter[2], C. Koos[1], W. Freude[1], J. Leuthold[1,5]

[1] Institutes IPQ and IMT, Karlsruhe Institute of Technology (KIT), Karlsruhe, Germany

[2] Rainbow Photonics AG, Zurich, Switzerland

[3] Photonics Research Group, Ghent University – IMEC, Department of Information Technology, Gent, Belgium

[4] IMEC, Leuven, Belgium

[5] Institute of Electromagnetic Fields, ETH Zurich, Zurich, Switzerland

Silicon waveguides can be functionalized with an organic $\chi^{(2)}$-nonlinear cladding. This complements silicon photonics with the electro-optic (EO) effect originating from the cladding and enables functionalities such as pure phase modulation, parametric amplification, or THz-wave generation. Claddings based on a polymer matrix containing chromophores have been introduced and their strong $\chi^{(2)}$-nonlinearity has already been used to demonstrate ultra-low power consuming modulators. However, these silicon-organic hybrid (SOH) devices inherit not only the advantageous properties. Said polymer claddings require an alignment procedure called poling and must be operated well below their glass transition temperature. This excludes some applications. In contrast, claddings made from organic crystals come with a different set of properties. In particular there is no need for poling. This new class of claddings also promises a stronger resilience to high temperatures, better long-term stability and photo-chemical stability. We report on the deposition of an organic crystal cladding of BNA (N-benzyl-2-methyl-4-nitroaniline) on silicon-on-insulator (SOI) waveguides, which have a CMOS-like metal stack on top. Adhering to such an architecture, which preserves the principal advantage of using CMOS-based silicon photonic fabrication processes, permits the first demonstration of high-speed modulation at 12.5 Gbit/s in this material class, which proves the availability of the EO effect from BNA on SOI also for other applications.

4.1.1 Introduction

The telecommunications industry heavily relies on modulators based on the $\chi^{(2)}$ linear electro-optic (EO) effect. Modulators exploiting this effect are popular because they allow for reliable phase and amplitude encoding of the most intricate modulation formats [105] in any desirable pulse shape [106] and up to highest speed. The general requirements on such modulators are demanding. They should offer lowest drive voltages (V_π) on a small footprint. They should be mass producible, and offer reliable operation under any common environmental condition.

State-of-the art modulators are based on $LiNbO_3$ and meet the above criteria – except that they have a fairly large footprint. They have recently been challenged by integrated GaAs modulators [C5] that are more compact. However, industry is interested in silicon photonics as a platform for integrated optics. Making silicon based devices allows resorting to an extensive infrastructure and fabrication experience from the CMOS electronics industry with lithographic resolution for a feature size down to 22 nm. Furthermore, one potentially could fabricate devices more economically. In spite of the many advantages of silicon, this crystalline material does not possess a $\chi^{(2)}$-nonlinearity due to its centro-symmetric structure. There are other options to construct phase modulators with silicon waveguides, e.g. by using the plasma dispersion effect in silicon [24], [107], [108]. For many applications this presents a simple and effective solution. However, the plasma dispersion effect is inherently related to the plasma absorption effect. Thus, due to the absorption, it can be challenging to precisely access particular points in complex QAM con-constellation diagrams without pre-distortion of the electrical driving signal.

In addition, new applications based on optical parametric amplification, second harmonic generation for frequency conversion, or mid-IR applications can be implemented using a $\chi^{(2)}$ nonlinearity [109]. The Kerr effect based on a $\chi^{(3)}$ nonlinearity may be used instead of a $\chi^{(2)}$-nonlinearity in some cases. Yet, $\chi^{(3)}$ nonlinearities typically require significantly higher threshold powers [J11]. To obtain a $\chi^{(2)}$-nonlinear effect in silicon, strain has been applied, thereby breaking the lattice symmetry [110]–[112]. However, larger nonlinearities can be achieved by employing nonlinear organic materials on the silicon platform. By exploiting the EO effect from chromophores hosted in polymers [35], [113], this so-called silicon-organic hybrid (SOH) approach [8] has already been demonstrated for 40 Gbit/s high-speed phase modulation [J14], or 112 Gbit/s IQ modulation using a 16QAM [J4] format. Modulators based on chromophores can excel with very low V_π voltages, resulting in ultra-low power modulators [J5]. Unfortunately, the approach with chromophores hosted in polymers has issues on its own. First, operation is limited to be well below the glass transition temperature of the organic material. Second, these devices require an additional fabrication step, in which the chromophore molecules need to be aligned by an external electrical field at each modulator at elevated temperatures. This step is commonly referred to as poling of the nonlinear material and presents an additional effort, especially when producing arrays of modulators in dense photonic integrated circuits (PIC) [114].

The disadvantages due to the use of chromophores hosted in polymers can be overcome by substituting them with a new class of $\chi^{(2)}$-nonlinear materials based on organic crystals. Organic crystals are particularly attractive, as one can choose from a large variety. Organic crystals can be designed to have high melting temperatures and good photo-chemical stability. Therefore they can withstand harsh conditions. No poling procedure is needed. Some of these crystals are already commercialized for THz-wave generation when irradiated by high-power laser pulses with femtosecond to nanosecond duration [115]. However, special methods have to be developed in order to envelop sub-micrometer silicon waveguide structures with a solid crystal. A low-speed phase shifter made of organic single crystals has been demonstrated [116]. Other successful demonstrations include the testing of a horizontally slotted silicon waveguide filled with such an organic crystal [117], or a tunable organic-crystal micro resonator [118].

In this paper, we demonstrate high-speed EO intensity modulation at 12.5 Gbit/s in a Mach-Zehnder modulator (MZM) employing an organic crystal. We use BNA [119] as a source for the linear electro-optic effect, describe the growth of this organic crystal on silicon waveguides, and quantify the resulting $\chi^{(2)}$-nonlinearity. Fabrication has been done with metal electrode stacks similar to those of standard CMOS electronic circuits in order to maintain CMOS compatibility before the deposition of organic material. Growing an organic crystal on such a silicon photonic chip surface is of particular importance as the CMOS stack permits crossings between optical waveguides and electrical transmission lines. The modulation speed is not limited by the crystal itself, but rather by the electrical transmission line.

4.1.2 SOH Concept Employing Organic Crystals

The structure of an SOH phase-shifter and the MZI configuration used to achieve intensity modulation is depicted in Fig. 4.1. An organic cladding material is placed to cover an optical waveguide such that the organic material interacts with the optical field of the SOI waveguide (WG). A particular strong overlap [J6] of the guided light with the organic cladding is obtained with the so-called strip-loaded slot WG structure shown in Fig. 4.1(a). In this structure the optical field is enhanced in the slot as the refractive index is lower in the slot than in the silicon rails, see Fig. 4.1(b). The confinement of light to the organic material is best for a quasi-TE mode, where the dominant electric field component is in x-direction, i.e., parallel to the substrate plane, Fig. 4.1(b). To induce a phase-shift by means of the $\chi^{(2)}$-nonlinear effect, an RF voltage must be applied across the slot, Fig. 4.1(c), for controlling the refractive index of the cladding. Switching is fast, if the strip-loads and rails are sufficiently doped in order to guarantee a good conductivity and therefore allow a fast charging and discharging of the slot capacitance. However, excessive doping leads to excessive optical losses. The two silicon rails are electrically connected by the silicon strip-loads to metal electrodes far away from the optical field in the slot, Fig. 4.1(d). The $\chi^{(2)}$-nonlinear refractive index change is most efficient, if the largest organic cladding's diagonal $\chi^{(2)}$-tensor element (green arrows in Fig. 4.1(a)) is aligned along the x-axis perpendicularly to the slot sidewalls and thereby in the direction of the modulating RF field $E_{x,\mathrm{RF}}$ (red arrows). In a bulk material with EO coefficient r and a plane wave (wavelength λ) one then can expect a phase shift in the order $\Delta\varphi = \Gamma(\pi/\lambda)Ln^3 rE_{x,\mathrm{RF}}$ for a device of length L in a homogeneous material with refractive index n and a confinement Γ of the optical mode in the slot. The product $n^3 r$ is used as an EO figure of merit (FOM) to describe the ability of an EO material to shift the phase of an optical wave upon application of a modulating voltage.

For intensity modulation the phase-shifters are typically arranged in the two arms of an MZI, Fig. 4.1(d). The phase shift in each arm is controlled by the ground-signal-ground electrodes of a coplanar transmission line [120], [121]. Since the electro-optic material is arranged along one direction only, the optical fields in the two arms experience phase shifts in opposite direction when they are operated with the GSG electrodes. This is called push-pull operation mode of the MZM.

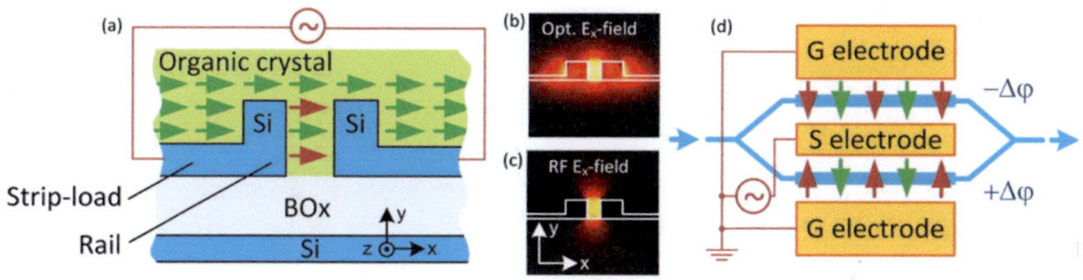

Fig. 4.1: SOH phase shifter waveguide (WG) with organic crystal cladding and its use in a Mach-Zehnder modulator (MZM). (a) Cross section of strip-loaded slot WG to be used with light in quasi-TE mode. Strongest diagonal $\chi^{(2)}$-tensor element of organic crystal cladding (polar axis indicated by green arrows) in the direction of the applied RF field (red arrows). (b) Cross section of color-coded dominant x-component $|E_x|$ of the optical E-field in quasi-TE mode in the strip-loaded slot WG. (c) Electrical RF field in x-direction (red arrows in subfigures). (d) Mach-Zehnder interferometer modulator (MZM) with optical waveguides in blue, electrical connections in red, ground-signal-ground electrodes (GSG) of coplanar transmission line in yellow. In each arm of the interferometer a phase modulator section is inserted. It consists of a strip-loaded slot WG. Note that the organic crystal has the same orientation in both arms, such that an electrical field applied to the central signal electrode (S) will introduce opposite phase shifts. This mode of operation is called "push-pull" and allows pure amplitude or intensity modulation without any phase modulation.

To apply an organic EO cladding we have two choices:

1. One may apply polymers which host chromophores [J7], [J14]. This is frequently done because of the large electro-optic coefficient r and resulting large FOM. However, the chromophores need a one-time alignment ("poling") in order to develop a macroscopic EO effect. Also, there might be an issue when using the polymers at elevated temperatures, because the chromophores might slowly lose alignment even before the polymer's glass transition temperature is reached.
2. In contrast, the material class comprising organic EO crystals offers the advantage that the chromophores composing these crystals keep their alignment as determined by the crystal structure up to the crystal melting temperature (provided the molecules remain chemically stable). Hence, using materials from the class of organic crystals promises long-term stability even at elevated temperatures and high illumination intensities due to their superior photo-chemical stability [122].

An overview of available materials of this class is given in [123], p. 163. in [124]. A few examples are listed here for reference: Stilbazolium salt crystals like DAST (4-N, N-dimethylamino-4'-N'-methyl-stilbazolium tosylate) with $n^3r = 480$ pm/V melting at 256 °C [125], DAPSH (trans-4-dimethylamino-N-phenyl-stilbazolium hexafluorophosphate) with $n^3r > 500$ pm/V, and DSTMS (4-N,N-dimethylamino-4'-N'-methyl-stilbazolium 2,4,6-trimethylbenzene-sulfonate) with $n^3r = 480$ pm/V melting at 258°C [126]. However, DAST, DAPSH and DSTMS are not stable at melting temperatures, so only solution growth is possible [123]. Furthermore, configurationally locked polyene molecular crystals have been developed [109], which are stable upon melting, which makes melt growth possible. These materials deliver similarly high effective nonlinearities. Examples are OH1 (2-(3-(4-hydroxystyryl)-5,5-dimethylcyclo-hex-2-enylidene)malononitrile) $n^3r = 470$ pm/V melting at 212 °C, or OH2 with an expected FOM of $n^3r > 600$ pm/V melting at 242 °C.

Here, we use an organic crystal made from BNA [119]. It has a high diagonal nonlinear optical coefficient $d_{333} = 234$ pm/V at 1064 nm [119] and a relatively low melting temperature of 105 °C. Nevertheless, we chose BNA for our proof-of-principle experiments, because it was easy to grow due to the low melting point and its reasonable wetting properties on a CMOS-structured silicon chip.

4.1.3 SOH Device Fabrication and Experiments

Mach-Zehnder modulators with SOH waveguides comprising the organic BNA crystal as a cladding and standard CMOS metal stack electrodes have been fabricated. After structuring the SOI wafer, we cover the SOI WGs with a thin film of the organic crystal BNA and open areas for electrical contacting to supply the modulating voltage. To verify and quantify the resulting EO effect, we use the modulator to encode data on an optical carrier.

Fabrication of an SOH Modulator

In Fig. 4.2 the cross section of an MZM consisting of two strip-loaded slot WGs is shown, in our implementation with a CMOS-like metal stack. The organic crystal BNA fills the WG slots. Other chips from the same wafer have been used before to make EO modulators based on EO-active chromophores hosted in a polymer [J4].

The fabrication of the passive part follows Ref. [J4]: SOI wafers (SOITEC) with a WG layer (220 nm high) on a buried oxide (Box, 2 µm thick) are processed employing 193 nm deep UV lithography at IMEC. Slot WGs ($w_{\text{Slot}} = 125$ nm, $w_{\text{Rail}} = 220$ nm, $h_{\text{Rail}} = 220$ nm) with attached n-doped silicon striploads ($h_{\text{Stripload}} = 50$ nm, arsenic doping with a nominal concentration of $3\times10^{17}\,\text{cm}^{-3}$) are etched into the WG layer. Standard strip and rib WGs for low-loss access waveguides [38] and standard grating couplers [40] are structured by dry etching 70 nm of Si.

The copper electrodes (coplanar RF transmission lines guiding the electrical modulating wave) and the strip-loaded slot WG (guiding the light) are connected by conducting tungsten-filled (W) vias and a silicide film (surrounded by highly doped silicon with a concentration of nominally $1\times10^{20}\,\text{cm}^{-3}$). Dielectric layers of mostly SiO_2 and Si_3N_4 support the Cu electrodes and fix the distance between Cu electrode and BOx to $d_{\text{Cu-BOx}} = 1.1$ µm. This metal stack could be extended with additional standard layers, e. g., making aluminum pads for packaging. However, in our case we end the stack with a thin layer of SiC to protect Cu from air. The slot WGs are 1.5 mm long and terminated at each end with a transition from the slot WG to standard strip WGs using low-loss converters [J8]. Multi-mode interference (MMI) couplers split the input field and combine the fields in the two interferometer arms.

The trenches to expose the slots are opened by dry and wet etching of the dielectric layers, which are then filled with BNA that in turn is covered with a layer of polyvinyl alcohol (PVA).

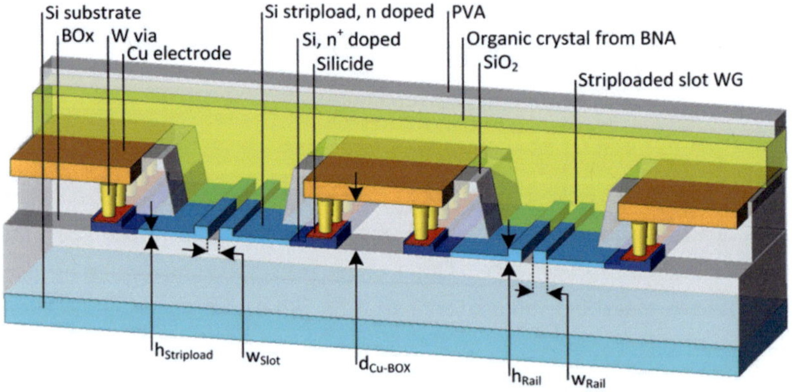

Fig. 4.2: Cross section of MZM employing an organic crystal (not drawn to scale). Two phase modulator sections are displayed consisting of strip-loaded slot WGs, which are filled with a nonlinear organic BNA cladding. The silicon rails are connected to ground-signal-ground electrodes of a coplanar transmission line through the Si strip-loads, a silicide layer and tungsten vias. This electrode arrangement allows crossings of optical WGs and electrical transmission lines. This architecture corresponds to the metal stack technology well known from the CMOS platform.

Deposition of Organic Crystal BNA

In integrated optics, there are various possibilities to fabricate claddings of organic single-crystal structures [123], [127]. In the past, bulk crystals of DAST have been grown and placed on top of silicon waveguides [125]. For this an elaborated polishing procedure is required, and obtaining a good optical contact between a crystal and an SOI chip surface is a challenge. Moreover, the thickness of the crystal might be inconvenient when attempting to contact modulation electrodes. Another method relies on growing single-crystalline thin films of various crystals such as DAST and DSTMS from a solution between glass plates. However, transferring the thin films to the chip implies the risk of fractures. Thus, for processing a full wafer, on-chip crystal growth seems to be the only viable option. For chips with very smooth topography, single-crystalline films of OH1 have been successfully grown directly from solutions on SOI chips [123], which takes several weeks. The most versatile technique, however, is to fabricate organic single-crystalline films by using melt growth. Using this approach, thin organic single-crystalline wires with a thickness down to 25 nm have already been grown on glass [128], and also more complex waveguiding structures such as microring resonators were fabricated [118].

Here we opt for direct, on-chip growth of BNA from the melt. This approach is not impeded by the rugged surface (due to the CMOS metal stack), shows reasonably large growth rates (hours), and provides a high-purity crystal. The crystalline symmetry of BNA is orthorhombic *mm2*. Its melting temperature is relatively low, about 105 °C, which is the limiting temperature for post-processing possibilities. BNA thin films always grow along the *ac* crystallographic plane with the fastest growth along the *a* direction, i. e., normal to the polar axis *c* [117], which we confirmed by polarized second-harmonic generation experiments. Therefore, when we manage to induce the growth of BNA crystals along the WG direction, its largest $\chi^{(2)}$-tensor element $\chi^{(2)}_{333}$ coincides with the E_x-field of the quasi-TE mode in the slot WGs.

We employ the following procedure to grow BNA films:

(1) Place a flat glass plate (a 200 µm thick borosilicate glass wafer) covering the target area on the Si chip. Deposit the BNA as powder on one side perpendicularly to the phase modulator sections of the WGs as shown in Fig. 4.3(a).

(2) In vacuum, heat the whole sample up to 120°C, well above the melting temperature of BNA to decrease the viscosity of liquid BNA. The capillary effect between SOI chip and glass plate pulls the liquid BNA below the glass plate, see Fig. 4.3(b). The wetting properties for BNA on both SOI and glass are sufficient to obtain a nice capillary flow. In case of other materials special surface treatment might be necessary. The vacuum helps to prevent voids when filling.

(3) After a few minutes, the melt is distributed homogeneously below the glass plate. Then the vacuum is replaced by nitrogen at normal atmospheric pressure, and the sample is quickly cooled down to room temperature. This leads to fast

crystallization of BNA. However, the resulting film is polycrystalline due to its fast growth well below the melting temperature.

(4) To get the same single-crystal orientation in all the phase modulator sections, a controlled crystallization is necessary. By heating the entire chip covered with polycrystalline BNA to just below the melting temperature (100 °C) and by keeping one side at a fixed temperature above the melting point (105°C), the polycrystal melts again, but only at the "hot" side.

(5) When removing the additional, one-sided heat source, the temperature of the "hot" side reduces back to 100°C. Starting with a random orientation at the "cold" side, the crystal then grows along its fastest growth direction which is oriented towards the formerly "hot" side, Fig. 4.3(c). Note that spontaneous nucleation of BNA does not occur at this temperature unless cooled down well below the melt temperature, as done in step (3). Hence it is assured that the film only starts growing from the "cold" top (seed) area.

(6) To access the RF electrodes the glass plate is removed as shown in Fig. 4.3(d), and the entire chip is spin-coated with polyvinyl alcohol (PVA) to reduce scattering from potential crystal fractures induced by removal of the glass plate.

The direction of the optical axis of the BNA crystal was determined with a reflection microscope by placing and rotating the sample between crossed polarizers, thereby measuring the crystal's birefringence. In different deposition attempts we achieved a crystalline orientation deviating from the optimal direction of the polar axis by $\alpha = 10°\ldots40°$. Additional growth-guiding trenches on chip or in the auxiliary glass plate could optimize the orientation further (for well-defined micro-sized channels the growth direction aligns perfectly [117], [118], [128]). We manually removed the organic crystal locally for contacting the electrodes with RF probes. In practice, lithographic processes [116] for structuring organic crystals will be preferred.

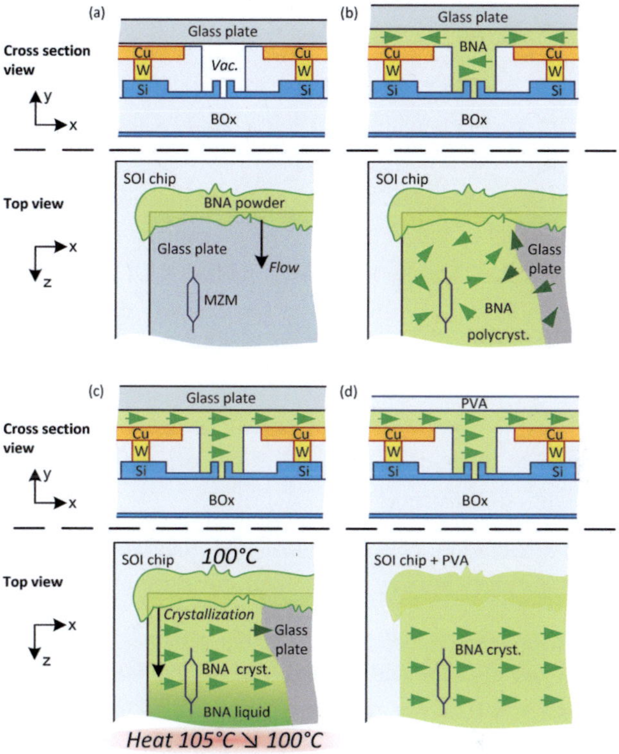

Fig. 4.3: Deposition method for growth of an organic single crystal thin film on top of strip-loaded slot WGs. Upper row: Cross-section of one phase modulator section as in Fig. 4.2. Subfigures (a)…(d) show various stages of the deposition process. BNA (green shading) covers the chip and fills the slot beneath an auxiliary glass plate (grey shading). Lower row: Tow view of chip (light grey shading) with auxiliary glass plate on top (grey shading). BNA (green shading) fills the space above the chip and below the auxiliary glass plate. The deposition starts by putting BNA powder at one edge of the auxiliary glass plate. More precisely: (a) A vacuum is maintained between glass plate and chip. (b) Applying heat melts the BNA powder, which is drawn into the gap between glass plate and chip due to the capillary effect. Polycrystalline BNA forms after a fast cool-down process. No macroscopic $\chi^{(2)}$-nonlinearity is to be seen. Green arrows indicate the random orientation of the polar axis of the polycrystalline domains. (c) Region of applied BNA power is heated to a fixed temperature of 100 °C close to but below the melting point ("cold" side). An additional local heat source increases the temperature of the opposite ("hot") side to a temperature of 105 °C, which is above the melting point. When removing the additional local heat source, the temperature of the hot side is reducing back to 100 °C. Beginning with a random orientation at the cold side, a single crystal then grows along the fastest growth direction which is oriented towards the formerly hot side. (d) After removal of the auxiliary glass plate, a PVA cover is spin-coated to fill any cracks in the BNA for reducing scattering in the WGs.

4.1.4 SOH Modulator for Data Generation

The viability of the EO modulation concept and successful deposition of BNA is demonstrated in an experiment. Fig. 4.4(a) depicts the setup. A 12.5 Gbit/s electrical data signal with a non-return-to-zero pseudo-random binary sequence (PRBS) of length $2^{31}-1$ is generated and amplified to nominally $V_{pp} = 8$ V. RF probes bring this signal to the SOH MZM and terminate the electrodes with an off-chip resistor of 50 Ω. Two bias-Ts allow adding a DC voltage for adjusting the MZM to its quadrature operating point for intensity modulation. Light with a wavelength of 1538 nm from a tunable laser source (TLS) is amplified in an erbium-doped fiber amplifier (EDFA) to a power of 18 dBm, and coupled via a grating coupler (GC) to the MZM. The resulting on-off keyed (OOK) optical signal is collected at the output GC, amplified, filtered, and received with a digital communication analyzer (DCA).

In this first generation chip we measured relatively high fiber-to-fiber losses. Potential loss sources have been identified. They are as follows:

(1) Coupling losses of about 5 dB for each GC. In a commercial implementation more effort would be made to improve this loss. More elaborate GCs exhibit losses of 1.6 dB per coupler [129]. Inverted tapers for butt coupling of external fibers were demonstrated to have a loss of 0.7 dB [130].

(2) Losses of about 5 dB/cm in the 7 mm long access WGs, which have bends and strip-to-ridge waveguide transitions. Further we have losses of about 1.5 dB in strip waveguides by scattering due to WG sidewall roughness and by absorption due to the crossing metal electrodes.

(3) Excess loss of about 1 dB for each of the two MMIs adds to the splitting loss of 3 dB.

(4) Loss of about 1 dB in each of the two strip-to-slot transitions.

(5) Loss in the phase modulator section of about 16 dB. A reduction of losses seems to be relatively straight-forward for all issues observed in (1)-(4), so that device insertion loss effectively is only limited by the loss in the phase modulator section. The slot WG sidewall roughness could be improved by using atomic layer deposition [J13], which at the same time could enhance the wetting properties for organic crystal growth.

Further, variations in the fabrication process have led to an imbalance of the loss in both arms of the MZM. An extinction ratio of more than 6 dB was measured with modulation, as can be seen from the eye diagram Fig. 4.4(b). The 3 dB-bandwidth of the MZM was measured to be about 7 GHz. It is limited by the RC constant of the structure and RF wave propagation losses [J4]. The V_π-voltage at DC is around 8 V leading to a $V_\pi L$ product of 12 Vmm. This corresponds to a figure of merit (FOM) for the EO coefficient of $n^3 r = 31$ pm/V. Evaluation of the open eye diagram in Fig. 4.4(b) of the received optical signal indicates a signal quality Q^2 of 14 dB, which translates into a bit error ratio (BER) of 3×10^{-7}. From the nonlinear coefficients d_{333}, d_{311} and refractive indices reported in [22] and using the same model as in [28] we estimate the EO coefficients of the organic crystal BNA to be $r_{33} = 24$ pm/V and $r_{13} \sim r_{51}$ to be negligible. The expected material's FOM should be in the order of $n^3{}_3 r_{33} = 135$ pm/V. Taking the measured crystal misalignment of $\alpha = 34°$ of the device under test into account, we would expect a FOM of $n^3{}_\alpha r_{33} \cos^3\alpha = 67$ pm/V. However, with a value of 31 pm/V we only find a value that is about half of the expected nonlinearity.

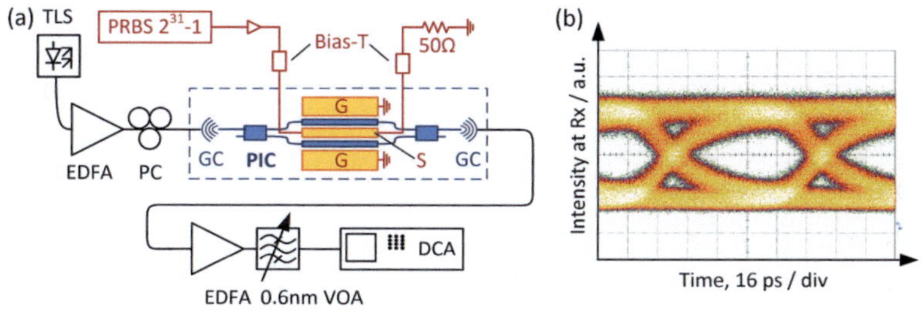

Fig. 4.4: Experimental setup for demonstrating the EO effect of a hybridly integrated organic crystal. Optical on-chip connections are in blue, electrical connections are drawn in red. (a) CW light from a tunable laser source (TLS) is amplified by an erbium-doped fiber amplifier (EDFA), polarization-controlled (PC), and coupled by a grating coupler (GC) to one input of a multimode interference (MMI) coupler. The MMI maps the light onto the two arms of the Mach-Zehnder interferometer modulator with phase modulators in its arms as discussed in Fig. 4.2. Another MMI recombines the fields. This light is coupled back to a fiber by means of a GC and guided to the receiver. The receiver consists of an EDFA, a filter and a digital communication analyzer (DCA). A pseudo-random binary sequence (PRBS) is amplified, and passed through a bias-T for adjusting the MZM to the quadrature operating point. Ground-signal-ground (GSG) electrodes excite an electrical traveling wave, which modulates the incoming light, which is evaluated in the receiver. (b) Eye diagram of the received optical signal at 12.5 Gbit/s. A quality factor Q^2 corresponding to 14 dB evaluates to a bit error ratio (BER) of 3×10^{-7}.

As a reason for this discrepancy, we suspect either imperfect filling of the slot WG (as the wet-etch step might have partly under-etched the WG and thus might have produced voids with imperfect crystal growth), or an overestimation of the original material's EO coefficient.

4.1.5 Conclusions

We demonstrated data modulation at 12.5 Gbit/s using an organic crystal of BNA integrated on SOI strip-loaded slot waveguides in a Mach-Zehnder modulator configuration. We showed that an important class of claddings, namely organic EO crystals, is compatible with the CMOS-typical metal stack used for traveling wave electrodes. Compared to frequently investigated, polymer-based EO claddings, our technique represents a viable alternative for fabricating high-speed modulators for advanced modulation formats. Organic EO crystal claddings pave the way for silicon-organic hybrid modulators with higher operating temperatures and better overall stability. This allows running the devices at larger optical input powers. Due to the strong light confinement, exceptionally large intensities can easily be reached, which might turn out to be useful not only for modulators but also for efficient THz-wave generation and for parametric amplification.

[End of Paper]

4.2 IQ Modulator Based on Chromophores in a Polymer

[Start of Paper, *Opt. Express*, vol. 21, no. 11, pp. 13219–13227, Jun. 2013]

Silicon-organic hybrid (SOH) IQ modulator using the linear electro-optic effect for transmitting 16QAM at 112 Gbit/s

Dietmar Korn[1,*], Robert Palmer[1], Hui Yu[2], Philipp C. Schindler[1], Luca Alloatti[1], Moritz Baier[1], René Schmogrow[1], Wim Bogaerts[2], Shankar Kumar Selvaraja[2], Guy Lepage[3], Marianna Pantouvaki[3], Johan M.D. Wouters[3], Peter Verheyen[3], Joris Van Campenhout[3], Baoquan Chen[4], Roel Baets[2], Philippe Absil[3], Raluca Dinu[4], Christian Koos[1], Wolfgang Freude[1], and Juerg Leuthold[1,5]

[1] Karlsruhe Institute of Technology (KIT), Institutes IPQ and IMT, Karlsruhe, Germany

[2] Photonics Research Group, Ghent University – IMEC, Department of Information Technology, Gent, Belgium

[3] IMEC, Leuven, Belgium

[4] GigOptix Inc., Bothell (WA), USA

[5] Now with ETH-Zurich, Zurich, Switzerland

Advanced modulation formats call for suitable IQ modulators. Using the silicon-on-insulator (SOI) platform we exploit the linear electro-optic effect by functionalizing a photonic integrated circuit with an organic $\chi^{(2)}$-nonlinear cladding. We demonstrate that this silicon-organic hybrid (SOH) technology allows the fabrication of IQ modulators for generating 16QAM signals with data rates up to 112 Gbit/s. To the best of our knowledge, this is the highest single-polarization data rate achieved so far with a silicon-integrated modulator. We found an energy consumption of 640 fJ/bit.

4.2.1 Introduction

Modulators that can reliably access any point within a constellation diagram are needed to encode signals with advanced modulation formats. The realization of these key components as photonic integrated circuits (PIC) on the silicon-on-insulator (SOI) platform holds promise for low power consumption, low cost and high volume production. Currently, $LiNbO_3$-based modulators are used for the most part. Exploiting the established infrastructure from scalable CMOS technology a new generation of silicon photonic devices emerges and is likely to substitute $LiNbO_3$, especially when arrays of modulators will be needed.

CMOS process compatibility for fabrication is essential to further silicon photonic modulators. A common approach is to confine the production of SOI modulators to a few simple steps such as silicon etching, doping & annealing, deposition of dielectric layers and metal electrodes. And indeed, the most common silicon modulators that are based on a pin or pn junction within a silicon ridge waveguide rely on these CMOS process steps. In these modulators free-carrier dispersion is employed by injecting [131] or depleting [108] carriers. This gives control over the phase of light, but also changes the absorption, which makes arbitrary waveform generation intricate. So far, numerous silicon modulators using this principle have been demonstrated in resonant configurations [132]–[135]. Also non-resonant designs [J10] were published showing an attractive bandwidth for on-off-keying (OOK) at data rates up to 50 Gbit/s [24]. While an increase in modulation bandwidth seems certainly possible [J14], bandwidth limitations in electronics would favor advanced modulation formats with reduced symbol rates. That means going to complex modulation formats like quadrature-phase shift keying (QPSK) as shown at 28 GBd in [136] to transmit 56 Gbit/s in a single polarization is a more advanced way to increase the bit-rate. In addition, polarization multiplexing can be added to further double the bit-rate. In the aforementioned publication [136] it has recently been shown how polarization multiplexing can be realized on-chip.

CMOS compatibility also sets limits to the available voltages. In light of rather high reverse bias and RF voltages reported for high-speed implementations (with respect to achievable phase shifts in silicon) it is advisable to also consider the linear electro-optic (Pockels) effect. The linear electro-optic effect can be found in strained silicon [111], [137]. Alternatively, the very common technique of spin-coating can be used to add an electro-optic, $\chi^{(2)}$-nonlinear organic cover layer on the modulator waveguide [138], [35], [139], [140] in order to create a silicon-organic hybrid (SOH) device. Applying a voltage then results in an instantaneous, pure phase shift, exactly as in $LiNbO_3$. This is an advantage over free-carrier based plasma effect modulation, where phase and amplitude modulation are linked. The free choice of cover ma-material brings the potential to reduce currently reported voltage-length-products for high-speed modulation to 3.8 Vmm (at 10 Gbit/s) [J5] or even lower for future advanced nonlinear organic materials, while pn-modulators so far show $V_\pi L \geq 10$ Vmm [141] (8.5 Vmm at 40 Gbit/s for a resonant structure [142]). The SOH approach combines the advantages of silicon (fabless development, fabrication infrastructure, scalability to high volume production) with the strong $\chi^{(2)}$-nonlinearity of an organic material.

In this paper we demonstrate the first IQ modulator for advanced modulation formats on the SOI-platform which is suited to transmit multilevel phase and amplitude encoded signals in the C-band. By applying the SOH concept, our approach with pure phase modulators in an interferometer structure gives us the freedom to choose any constellation, and enables arbitrary signal generation. Because of its relevance in applications, we decided for a 16QAM format for demonstrating the so far highest single-carrier single-polarization data rate of 112 Gbit/s on the silicon-platform. We further show error free generation and reception of a QPSK signal at 56 Gbit/s.

4.2.2 Structure of the Silicon-Organic Hybrid IQ Modulator

The IQ modulator is constructed by nesting two Mach-Zehnder modulators (MZMs) as shown in Fig. 4.1(a). These single-drive modulators are operated in push-pull mode at minimum transmission point, such that the resulting amplitude modulation of each provides the in-phase (I) and quadrature-(Q) phase component, when both MZMs are made to interfere with a phase shift of $\pi/2$.

To explain the SOH modulator concept a cross section of one MZM is presented in a simplified manner in Fig. 4.1(b). Light propagates in the two slot waveguides (WG) shown in blue. They constitute the arms of the MZM and are filled with the nonlinear material. The large index contrast between Si of $n_{\mathrm{Si}} = 3.48$ and the nonlinear polymer $n_{\mathrm{poly}} = 1.7$ causes an enhancement of the electrical field of the optical quasi-TE wave inside the slot, see Fig. 4.1(c). When a voltage is applied to the Si rails, it creates a strong electric field across the slot, see Fig. 4.1(d). Thus a large and therefore efficient overlap is obtained between the electrical and optical mode. By attaching thin Si striploads to the optical WG an electrical connection is made to an RF coplanar waveguide

(CPW), realized in a ground-signal-ground (GSG) configuration devised to have a 50 Ω impedance close to the RF source (similar to [57]). The active molecules (chromophores) of the nonlinear material are aligned during fabrication by applying a poling voltage (depicted in green) from one ground electrode to the next, which results in an orientation of the $\chi^{(2)}$-nonlinearity in both slots (green arrows) that is asymmetric with respect to the signal electrode (S). Thus in operation, when an RF signal on the S-electrode is applied, it will cause a positive phase shift in one arm and a negative one in the other, i.e., result in push-pull operation. Thus the MZM can deliver a pure amplitude modulation.

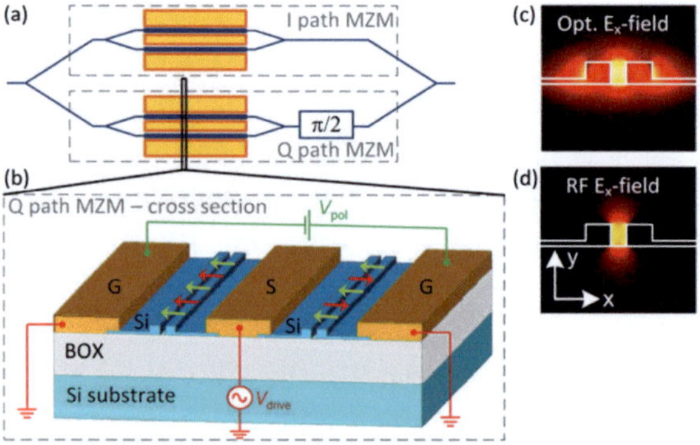

Fig. 4.1: IQ modulator based on the SOH concept. (a) Topview of the IQ modulator with nested Mach-Zehnder modulators (MZM), displaying optical waveguides (WG) in blue and electrical lines in orange. (b) Cross section of an SOH MZM, showing two silicon striploaded slot WGs, which act as phase shifters. They are filled and covered with a nonlinear cladding (not shown for clarity). The coplanar RF transmission line (GSG, ground-signal-ground) is impedance matched to the driving signal generator. The RF voltage at the S-electrode creates oppositely directed electric slot fields (red arrows). During the fabrication process, the $\chi^{(2)}$-nonlinearity is created by applying a poling voltage between both RF ground (G) electrodes at an elevated temperature. This aligns (poles) the active cladding molecules in a direction indicated by green arrows. In combination with the poled cladding, the modulating RF voltage leads to opposite phase shifts in both interferometer arms. (c) Color-coded dominant x-component $|E_x|$ of the optical electrical field in the slot WG cross section. (d) Modulating electrical RF field. Both fields are strongly confined to the slot, resulting in high modulation efficiency.

Our implementation of the MZM is depicted as a detailed cross section in Fig. 4.2. We start at IMEC with an SOI wafer (SOITEC). It has a 220 nm high waveguide (WG) layer on a 2 µm thick buried oxide (BOX). Using 193 nm deep UV lithography slot WGs ($w_{\text{Slot}} = 140$ nm, $w_{\text{Rail}} = 220$ nm, $h_{\text{Rail}} = 220$ nm) with n-doped silicon striploads ($h_{\text{Stripload}} = 50$ nm, As-doping with nominally 3×10^{17} cm^{-3}) are etched into the WG layer. Furthermore dry etching is employed to remove 70 nm of Si for standard grating couplers [40] and standard strip and rib WGs for low loss access waveguides [38].

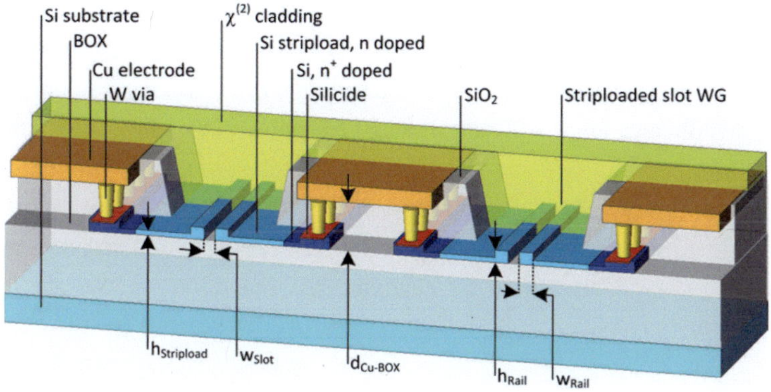

Fig. 4.2: Detailed cross section of MZM as implemented, showing two phase modulators with striploaded slot WGs, filled with nonlinear cladding; not to scale. Rails are connected to ground-signal-ground electrodes by tungsten vias, a silicide layer and the Si striploads. This electrode arrangement allows crossings of optical WGs and electrical transmission lines. Furthermore, it corresponds to the first part of standard metal stacks as known from CMOS technology.

A silicide film (surrounded by highly doped silicon with nominally 1×10^{20} cm^{-3}) connects through tungsten-filled (W) vias to the copper electrodes of the RF transmission line. Using this CMOS-like metal stack a conducting connection between the RF transmission lines (guiding the electrical modulating wave) and the rails (guiding the optical field) is established. It allows crossings of optical WGs with electrical transmission lines, where the surrounding dielectric layers of mostly SiO_2 and Si_3N_4 fix the distance to $d_{\text{Cu-BOX}} = 1.1$ µm. This metal stack is ready to be extended with additional standard layers, e.g. to make aluminum pads for packaging. We finished this metal stack with a thin layer of SiC to protect the Cu from air for this proof-of-principle device.

To transform the 1.5 mm long slot WGs into active modulator sections, trenches are etched into the dielectric layers to expose the slot WGs by a combination of dry and wet etching. A commercially available and reliable

[143] electro-optic polymer is spin-coated. This material (named M3 by the supplier GigOptix Inc. [114]) contains chromophores and is the very same material used in Telecordia certified polymer modulators of the same manufacturer. It is poled [144] inside the slot WG (alignment of the chromophores) by applying a DC voltage at elevated temperature to create the $\chi^{(2)}$-nonlinearity in the same way as in [J14].

The MZMs further consist of multi-mode interference (MMI) couplers. A transition from the slot WG to standard strip WGs is achieved by using a low-loss converter as described in [J8]. These single-drive modulators are in turn nested within one large MZ interferometer with a path length imbalance of 40 μm. The I and Q path are operated each in push-pull with ground-signal-ground (GSG) electrodes, such that their operation points can be set by applying a bias voltage along with the RF signal. The phase difference between I- and Q-component can be controlled by changing the operation wavelength in this proof-of-principle PIC.

4.2.3 Demonstrations

The performance of the IQ modulator is determined by the properties of its nested MZMs. The electro-optic small signal frequency response S_{21} of one MZM which is operated at its quadrature point is shown in Fig. 4.3. The 45 GHz RF probes were not de-embedded. The modulated light output power is detected with a photodiode. The raw frequency response of $|S_{21}|$ is recorded with a vector network analyzer (VNA). When switching off the optical carrier, a noise floor from the optical detector is seen. The blue curves result from a moving average applied to the measured data (red dots). The inset shows a blow up of the averaged frequency response in the low-frequency region. The receiver's equalizer (red curve) compensates the overall frequency response (magenta curve). The gray vertical line at 0.9 GHz marks the reference for normalizing the response function. The horizontal grey lines mark the -3 dB and -6 dB deviations from this reference point. The uncompensated -3 dB limiting frequency is 6.8 GHz, the -6 dB limit is 21 GHz.

This curve is atypical for a modulator insofar as the response drops sharply in a frequency range up to 1.7 GHz, see inset. Comparable curves were reported in Ref. [J14] for a silicon modulator with a very similar EO polymer (M1) from GigOptix. The skin effect becomes more pronounced for higher frequencies;

hence the RF loss increases strongly with frequency, especially for our relative-relatively thin electrodes. A 3 dB bandwidth of 6.8 GHz results for this modulator. For higher frequencies the response flattens resulting in a 6 dB bandwidth of 21 GHz. The region with the sharp sensitivity increase toward lower frequencies does not significantly affect the transmission quality as demonstrated in our experiments. Line coding and forward error correction tend to avoid the lower spectral regions anyway. Instead of equalizing at the receiver (red curve in Fig. 4.3) as used for QPSK, we used a pre-emphasis for 16QAM in the transmitter. We keep the pre-emphasis filter length short, which means that it can be implemented as a lookup table in the transmitter driver electronics, which is available anyway for generating multi-level signals for higher order modulation formats.

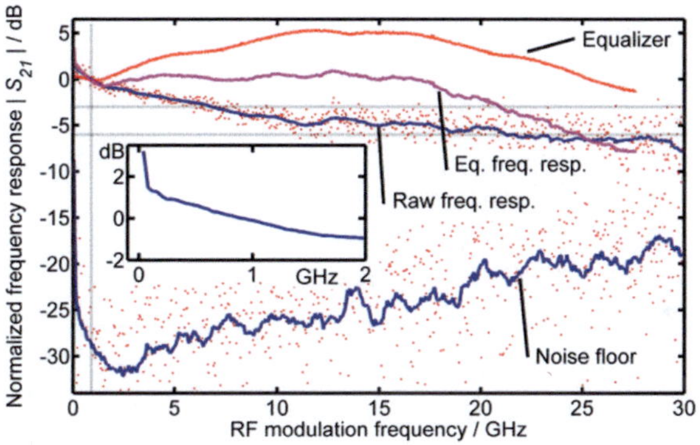

Fig. 4.3: Electro-optic frequency response S21 of our MZM (including the RF probes). The modulator is driven with a small-signal sinusoidal at the quadrature operating point. The modulated light output power is detected with a photodiode. The raw frequency response is recorded with a vector network analyzer (VNA). When switching off the optical carrier, a noise floor from the optical detector is seen. The blue curves result from a moving average applied to the measured data (red dots). The inset shows a blow up of the averaged frequency response in the low-frequency region. The receiver's equalizer (red curve) compensates the overall frequency response (magenta curve). The gray vertical line at 0.9 GHz marks the beginning of the frequency range which is of interest for data transmission (PRBS length 231-1). This frequency was chosen for normalizing the response function. The horizontal grey lines mark the -3 dB and -6 dB deviations from this reference point. The uncompensated -3 dB limiting frequency is 6.8 GHz, the -6 dB limit is 21 GHz.

To test data transmission with the SOH IQ modulator, two random signals with a pseudo-random binary sequence (PRBS) of length 2^{11}-1 have been created with an electrical arbitrary waveform generator (AWG) [145] at a symbol rate of 28 GBd (symbol duration is $T_s = 35.7$ ps). Our PRBS length was limited, but in [J7] we checked that a PRBS length of 2^{31}-1 applied to a comparable modulator structure led to comparable bit-error ratios (BER). After amplification to a peak-to-peak driving voltage of 5 V and having added bias voltages ($V_\pi = 2$ V at DC) of 0 V to 4 V (MZMs set to minimum transmission point), the electrical signal is fed via RF probes to the chip and connected to off-chip 50 Ω terminations, as shown in Fig. 4.4. Light at 1545 nm is coupled with grating couplers (GC, >10 dB for both couplers), and modulated in amplitude and phase. Before reception with an optical modulation analyzer (OMA) for error detection, the modulated light is amplified, filtered and attenuated as needed. The same setup is also employed to investigate the dependence of BER on the optical signal-to-noise ratio (OSNR) measured with an optical spectrum analyzer (OSA) while adding noise using an amplified spontaneous emission (ASE) source.

The device shows an extinction ratio of >26 dB. The measured overall optical insertion loss of 30 dB is high. However, we did not optimize all components for lowest loss. Our optical loss is composed of: (a) Coupling loss, which amounts to more than 10 dB for both grating couplers. Better grating couplers (with higher fabrication effort) promise coupling losses of 1.6 dB per coupler. (b) Loss in 7 mm long access WGs including bends and strip-to-ridge transitions (order of magnitude 5 dB/cm) by scattering (rough WG sidewalls) and absorption (WG partially located underneath metal).

The access WG could have been shortened to 1 mm, but the excess length facilitated our experiments. (c) Concatenation of four MMIs. (d) Strip-to-slot transitions, each contributing 1 dB loss due to fabricating tolerances. (e) Phase modulator section is estimated to have a loss of 10.5 dB. Optimizing (a)–(d) would reduce the loss essentially to the loss of the phase modulator section.

We summarize the basic devices properties: The active modulator section is 1.5 mm long, contributes 10.5 dB optical loss, has a V_π of 2 V at DC, and exhibits a 3 dB (6 dB) bandwidth of 6.8 GHz (21 GHz). The extinction ratio is larger than 26 dB.

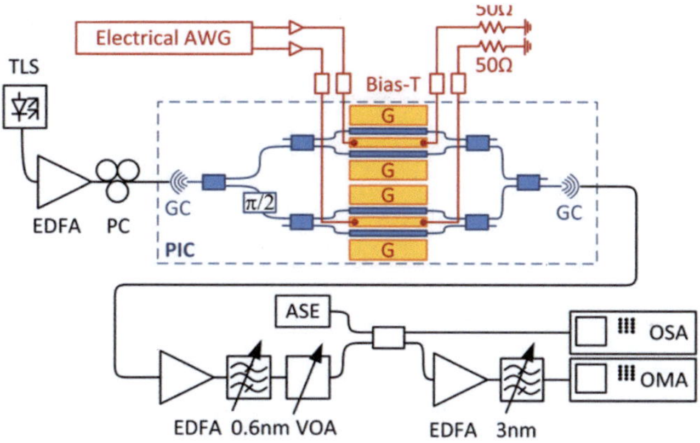

Fig. 4.4: Experimental setup with photonic integrated circuit (PIC) under test. In blue: Schematic SOI chip configuration with nested MZMs using MMIs, and with grating couplers (GC) to couple light at 1545 nm with cleaved standard single mode fibers (SSMF). In orange: Electric coplanar waveguides in ground-signal-ground (GSG) configuration contacted with RF probes to operate MZMs in push-pull mode. In red: Off-chip electrical components to supply the PRBS signal (electrical arbitrary waveform generator, AWG, electrical amplifier), bias-Ts (DC sources not shown) and termination. In black: Off-chip fiber based devices for characterization, including a tunable laser source (TLS), a polarization controller (PC), filters, erbium doped fiber amplifiers (EDFA), variable attenuator (VOA) and ASE source for OSNR tests.

QPSK, one of the most common formats used in coherent transmission systems is serving us as a benchmark. We can generate QPSK with the SOH modulators at a state-of-the-art symbol rate of 28 GBd, see Fig. 4.5. For the first time on the silicon platform we report, without relying on additional signal processing (such as pre-emphasis or equalization), a bit-error ratio (BER) of 4.5×10^{-4} for QPSK, i.e. well below the hard decision forward error correction threshold of 3×10^{-3}. The error vector magnitude (EVM, maximum normalization), a common measure for complex signals (directly related to the BER [146], [147]), is 24.9 % in this experiment. This measure provides a convenient tool for comparison with other, more advanced modulation formats, as used later. The imbalance between the I and Q path amounts to a factor of 1.3 and is due to unequal poling and imperfect adjustment of the operating points. By a careful adjustment of the I and Q voltages this imbalance can be compensated, and we did so for 16QAM.

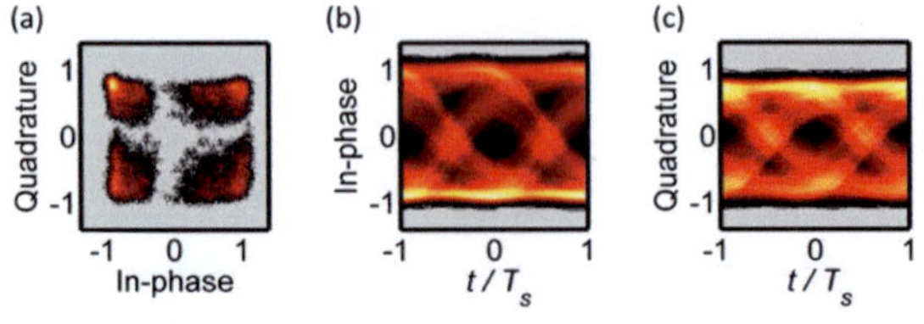

Fig. 4.5: QPSK generation at 56 Gbit/s with an IQ SOH modulator without using signal pre-emphasis, and detected without equalization. (a) Constellation and (b) in-phase and (c) quadrature-phase eye diagram. The BER is 4.5×10^{-4}, and the EVM is 24.9 % and thus well below the FEC error correction limit.

Generation of error free QPSK with equalization (19 taps, 1 per symbol) is reported next, see Fig. 4.6, for the first time on the SOI platform at 28 GBd, also for direct comparison with [136] using the same equalizer length to reach 56 Gbit/s (the highest data rate reported on one channel and polarization). The EVM is 14.2 % and mostly due to the path imbalance. No errors could be found over minutes. The BER is plotted over OSNR showing direct BER measurements and EVM measurements translated to BER. This modulator is 5.5 dB from the theoretical limit of QPSK at a BER of 3×10^{-3}. It is conceivable that equalization could be integrated on-chip along with driver electronics, as one can save on forward error correction efforts in this case.

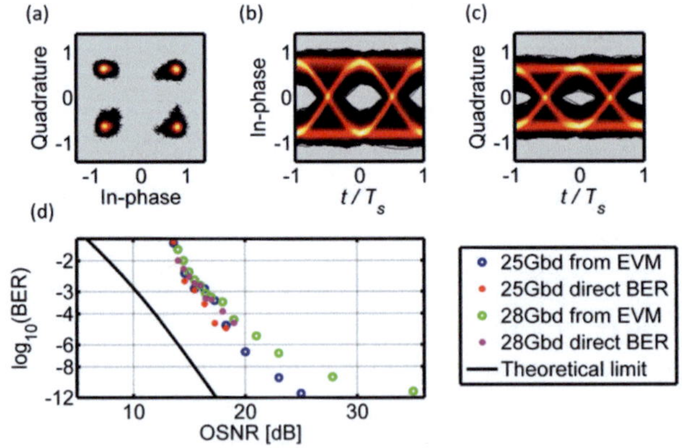

Fig. 4.6: QPSK generation at 56 Gbit/s with an IQ SOH modulator; detected using equalization with 19-taps as in [136]. Error free operation is measured and displayed in (a) constellation and (b, c) eye diagrams. The BER is shown in (d) in dependence of the OSNR at 25 GBd and 28 GBd. Measurements of the EVM are translated to BER and depicted.

16QAM is an advanced modulation format requiring complete control of ampli-amplitude and phase of the modulated signal. Using an 8-tap pre-emphasis, also correcting the IQ imbalance, the same modulator as above was used to generate an optical 16QAM signal at 28 GBd, i.e., 112 Gbit/s. This is currently the highest data rate on a single channel and polarization generated on the SOI platform, while still remaining below the FEC limit with a BER of 1.2×10^{-3}, an EVM of 10.3 %, without any equalization at the receiver, see Fig. 4.7.

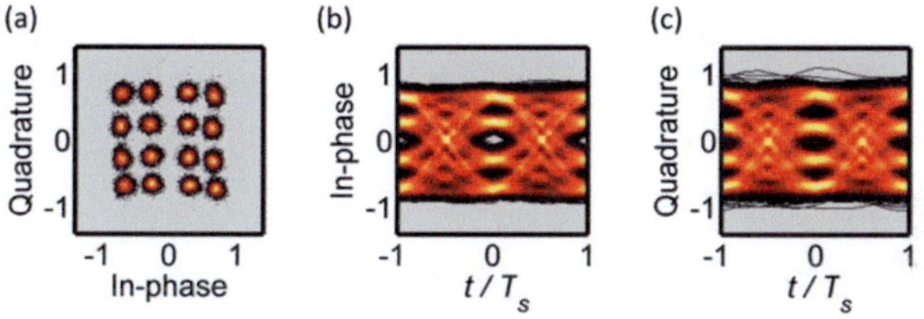

Fig. 4.7: Generation of 16QAM on a single channel and polarization at 112 Gbit/s with an IQ SOH modulator using pre-emphasis. (a) Constellation and (b, c) eye diagrams as observed when employing an 8-tap (1 tap per symbol) pre-emphasis at the transmitter, and no equalization at the receiver. The BER is 1.2×10^{-3}, and the EVM is 10.3 %.

4.2.4 Discussion and Conclusion

Advanced modulation formats not only bring higher spectral efficiency, but also a reduction of energy consumption of the modulator for a given driving voltage. Assuming a 50 Ω termination and a measured driving voltage of $V_{pp} = 5$ V (pre-emphasis reduces the effective voltage further, which is included in this estimation) we follow the recipe by [148] and find an energy consumption of 640 fJ/bit. Further improvements are to be expected when better nonlinear cladding materials are found [149] and, supposing driver electronics can be closely integrated, if the modulator's impedance can be designed to allow lower drive powers. Considering that an 8-tap pre-emphasis could be most easily implemented as a lookup table in the same electronics which generate the multi-level signal there is no demand for extra effort such as equalization.

This demonstration proves the potential of SOH modulators to compete with $LiNbO_3$ in its core domain of advanced modulation formats, and holds the promise for a reduction of cost, energy consumption, and size, in particular when making multichannel arrays. This extends the application range of silicon modulators further into long haul and access networks, which have been shown to strongly rely on ever more advanced modulation formats [22], while contend-contending with symbol rates in a very similar range as the one used in this experiment.

[End of Paper]

4.3 SiGe-Based Driver Circuit for SOH Modulators

It is important to consider electronic drivers for the SOH modulators for their deployment. While the production of electronic circuits on the same chip as the photonic structures is a long-term target, the compatibility of the fabrication processes is currently a major issue. Making separate chips with CMOS-based driver circuits is possible. However, mask costs for a dedicated driver design are much higher for a CMOS process than for making drivers with the SiGe platform, as indicated in a discussion with GigOptix.

SiGe-based drivers are favored for a (comparatively) low production volume. This is assumed to be reasonable for a market entry of SOH modulators, e.g., for long-haul applications. These applications require production volumes much lower than typical for consumer products. Here, we investigate the compatibility of SiGe drivers with SOH modulators with respect to the drive amplitude and the data rate (bandwidth).

Since SOH modulators have been shown to be useful at very low drive voltages, a commercial driver of GigOptix is mounted on a board customized (output impedance close to 50 Ω) to drive an SOH MZM of 1 mm length. The driver accepts electronic input signals with -20 dBm (50 Ω input, OOK signal), which alternatively could have come for instance directly from a photo detector. An electronic signal regeneration is performed and a 0 dBm (measured at an impedance of 50 Ω) output signal is sent directly, without an additional electrical amplifier, to the SOH modulator (a bias-T serves to set the operation point of the MZM to its quadrature point). The MZM is terminated with a 50 Ω load and driven at 12.5 GBd. A gate voltage of 100…300V (as described in Fig. 2.12) is applied. The π-voltages is ca. 0.8 V for the employed MZM. The modulator's insertion loss is roughly 25 dB. Below, a wide open eye diagram is displayed. EO modulation is demonstrated at high quality, i.e., a squared quality factor Q^2 corresponding to 11.7 dB, which corresponds to a bit-error-ratio (BER) of around 6×10^{-5}, Eq. (7.10). Apart from packaging issues this proves the viability of using SiGe circuits to operate SOH modulators. Instead of performing simple signal regeneration, also other functionality could have been implemented in the SiGe circuit.

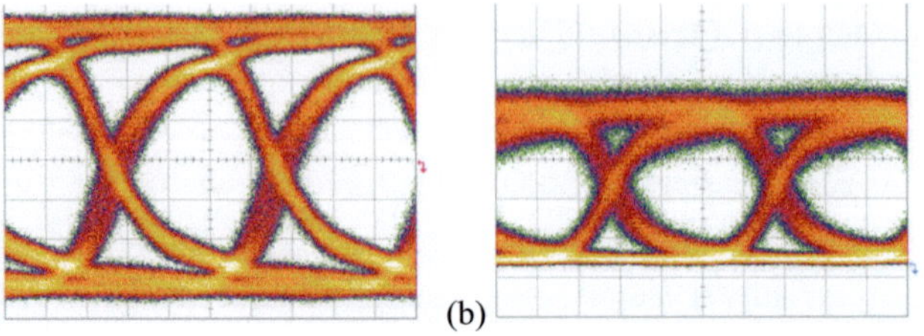

(a) (b)

Fig. 4.8: SOH MZM driven by SiGe driver circuit at 12.5 Gbit/s. Eye diagrams depicted with time on horizontal axis at 20 ps/div. (a) Electrical output with vertical axis at 100 mV/div of SiGe driver before being directly applied to the SOH MZM. (b) Received optical signal after SOH modulation, as recorded by a photodetector at 500μV/div. The squared quality factor Q^2 of the EO modulation corresponds to 11.7 dB.

5 Modulation with Benchmarking Devices

Benchmarking was done to understand the advantages of the SOH platform with respect to other platforms suited for the photonic integration of modulators. In this part of the thesis, we report on characterization results for SOI modulators using the plasma effect [J9], and modulators made on the GaAs platform [C5], [J3].

5.1 Plasma-Effect Based Device on Standard SOI Platform

This chapter has been published at the conference CLEO 2012 in San Jose, [C16]. © OSA 2012. The device under characterization was built as a modulator. We found that it can also be used for detection, i.e., a differential electrical current can be measured for incident, intensity modulated light. Other reports about defect-mediated sub-band-gap photo detection can be found in the literature, e.g., in [150]. This effect was further investigated in [J9].

[Start of Paper, *CLEO 2012*, p. CTu1A.1]

Detection or Modulation at 35 Gbit/s with a Standard CMOS-processed Optical Waveguide

Dietmar Korn[1], Hui Yu[2], David Hillerkuss[1], Luca Alloatti[1], Christoph Mattern[1], Katarzyna Komorowska[2], Wim Bogaerts[2], Roel Baets[2], Joris Van Campenhout[3], Peter Verheyen[3], Johan Wouters[3], Myriam Moelants[3], Philippe Absil[3], Christian Koos[1], Wolfgang Freude[1], Juerg Leuthold[1]

[1] Institute of Photonics and Quantum Electronics (IPQ) and Institute of Microstructure Technology (IMT),
Karlsruhe Institute of Technology (KIT), 76131 Karlsruhe, Germany

[2] Department of Information Technology, Ghent University-imec,
Center for Nano- and Biophotonics (NB Photonics),
St.-Pietersnieuwstraat 41, 9000 Ghent, Belgium

[3] imec, Kapeldreef 75, 3001 Leuven, Belgium

Light modulation and detection within a single SOI waveguide is demonstrated at 1550 nm. Multi-functional device concepts allow simplified transceiver systems and save processing steps, thereby increasing yield and reducing costs.

5.1.1 Introduction

Waveguide-based photo detectors and modulators are key components for highly integrated photonic circuits. Silicon Photonics is particularly appealing, because of the prospect to produce chips with existing CMOS infrastructure. Modulators [5] usually rely on free-carrier dispersion in all-silicon structures, whereas detectors either exploit defect states [151] or require deposition of Germanium [6] on silicon substrates. Uniting both functions on the same chip requires substantial processing effort and results in a decrease of yield.

In this paper, we demonstrate for the first time that a single device can be used to modulate and detect broadband communication signals with data rates of 35 Gbit/s. The device was fabricated using only the most common and reliable CMOS processes: Silicon dry etching and ion implantation. This significantly reduces the number of processing steps and allows for less complex fabrication runs with lower risk.

5.1.2 Device Design and Fabrication

Phase modulation is achieved by using carrier-depletion in a 3 mm long ridge waveguide which was doped symmetrically with P and N (both 2E18 cm-3, boron and phosphorus) dopants and annealed at 1075 °C for 10 s. The layout is shown in Fig. 5.1. Waveguides are defined on a standard p-doped SOI wafer (1E15 cm-3) with 2 µm buried oxide layer. Processing was performed in a standard CMOS line with 193 nm optical lithography. We start with a 220 nm high Si layer, use 70 nm dry-etch to define the ridge waveguides (500 nm width) together with shallow-etch grating couplers. 500 nm wide strip waveguides are then realized by fully etching the device layer around the phase modulation sections. Pt/Au coplanar waveguide electrodes were realized in Ground-Signal-Ground configuration (widths: 6 µm signal, ca. 100 µm ground, 3.5µm gap, pads larger but not shown). The electrodes are deposited on a BCB layer to enable crossing of optical waveguides and touch down only at strongly doped (silicide) P++ and N++ regions. These are connected by 400 nm wide and 1 µm long arms to the P-N-doped ridge waveguide. The doping was optimized for modulator applications.

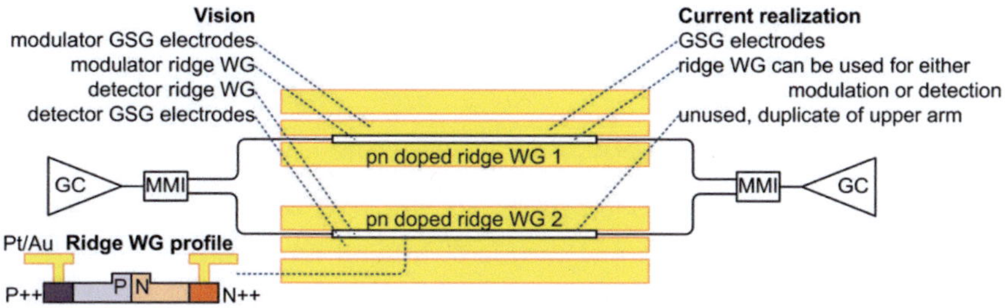

Fig. 5.1: Proposed and realized Mach-Zehnder Interferometer (MZI) on SOI, incorporating grating couplers (GC), multi-mode interference couplers (MMI), carrier-depletion ridge waveguides. The profile is depicted in the inset. WG 1 and WG 2 are identical; WG 1 was used for detection and modulation experiments. Vision: Use WG 1 for modulation, WG 2 for detection, e.g. for half-duplex transceiver or instant signal processing by using WG 2 to monitor and feedback to WG 1 (also packet or bitwise).

The operation principle of detection of sub-bandgap light in Si has been shown and explained by Knights et al. [152], relying on Si^+ implants to introduce defects associated with sub-bandgap states and P, N dopants (phosphorus, boron). Detection without Si^+ implants but weaker performance was shown by Zhao et al. [153]. Hence we use the same waveguide designed for phase modulation to detect light, suspecting incomplete annealing of P, N dopants to leave defects. In this first demonstration the focus has been on optimizing the quality of the modulator operation. In the future when more elaborated designs and electronics can be directly cointegrated on-chip, some disadvantageous detector performances may be optimized (e.g. the dark current could be made smaller, responsivity increased).

An imbalanced (40 µm) Mach-Zehnder Interferometer (MZI) configuration with multi-mode interference couplers (MMIs) was coupled to fibers via grating couplers (GCs, coupling loss 5 dB), see Fig. 5.1. The device allows for amplitude modulation and works over a wide wavelength range. We use only one arm to subsequently show modulation and detection, because of insufficient space to contact both coplanar electrical waveguides at the same time. However more functionality can be envisioned for this device: Imagine arm 1 of the MZI connected to driver electronics for modulation, arm 2 permanently attached to the receiver electronics on-chip. While such a device can act as a half-duplex transceiver, also signal monitoring and feedback for immediate signal processing are conceivable. Operating WG 2 at the same bias as WG 1 gives the best modulation extinction ratio, while a higher bias on WG 2 could be temporally used to increase receiver bandwidth and analyze the incoming signal.

5.1.3 Modulator and Detector Performance

Driving arm 1 at 5.7 V reverse bias with a non-return-to-zero (NRZ) pseudo random bit sequence (PRBS) of length 2^7-1 gives an on-off-keying (OOK) signal. The data stream is analyzed with a pre-amplified receiver, Digital-Communication-Analyzer (DCA) and Bit-Error-Probability Tester (BERT) and produces open eye diagrams shown in Fig. 5.2a. This proves error-free operation (BER < 2E-10) at 35 Gbit/s.

The detector's reverse bias is set to 7.3 V, close to the breakthrough voltage causing a dark current of 34 µA. Variation of the optical power at 14 dBm (in WG 1) gives a responsivity of 0.03 A/W. Using a signal created with a Lithium-Niobate (LN) modulator at 35 Gbit/s send at 14 dBm into WG1, a BER below 2E-6 is measured, well below the threshold for 2nd-generation forward-error correction (FEC), see Fig. 5.2b. Measurements of reverse bias current vs. optical input power (not shown here) exclude two-photon absorption as physical effect to explain sub-bandgap detection.

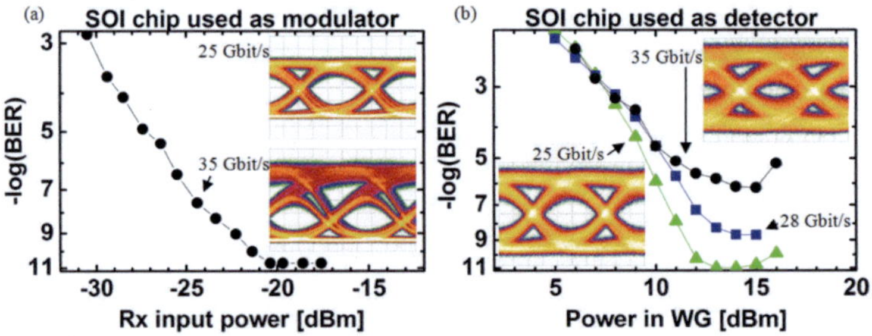

Fig. 5.2: Performance characterization around 1550 nm by data transmission of a PRBS (2^7-1) using NRZ OOK. (a) BER and open eye diagrams for modulation with SOI chip (reverse biased at 5.7 V) using commercial receiver. (b) BER and open eye diagrams for detection with SOI chip (reverse biased at 7.1 V) of optical signal (generated with LN modulator) using electronic amplifier.

5.1.4 Conclusion

Highspeed modulation or detection in the same SOI waveguide from fabrication relying only on dry etching and ion implantation has been demonstrated for the first time, allowing for multiple-purpose devices, e.g. instant signal processing with less effort. The striking extend to which silicon can be used for sub-bandgap light detection without dedicated fabrication steps is revealed, which allows for saving costs but also implicates for other detection concepts on SOI (e.g. plasmonics) to consider the influence on responsivity from silicon itself.

[End of Paper]

5.2 Electro-Optic Effect-Based Device on GaAs Platform

This chapter was published at OFC 2013 in Anaheim, USA [C5].

[Start of Paper, *OFC 2013*, p. PDP5C.4]

First Monolithic GaAs IQ Electro-optic Modulator, Demonstrated at 150 Gbit/s with 64-QAM

D. Korn[1], P. C. Schindler[1], C. Stamatiadis[2], M. F. O'Keefe[3], L. Stampoulidis[4], R. Schmogrow[1], P. Zakynthinos[5], R. Palmer[1], N. Cameron[3], Y. Zhou[3], R. G. Walker[3], E. Kehayas[4], I. Tomkos[5], L. Zimmermann[6], K. Petermann[2], W. Freude[1], C. Koos[1], J. Leuthold[1,7]

[1] Karlsruhe Institute of Technology (KIT), Institutes IPQ and IMT, Engesserstr. 5, 76131 Karlsruhe, Germany

[2] Joint Lab Silicon Photonics HFT4, Technische Universitaet Berlin, 10587 Berlin, Germany

[3] u2t Photonics UK Ltd, NETPark Incubator, Thomas Wright Way, Sedgefield, TS21 3FD, England

[4] Constelex Technology Enablers, Corallia Microelectronics Innovation Center, 15125 Athens, Greece

[5] Athens Information Technology Center, Peania-Attika 19002, Athens, Greece

[6] IHP, Im Technologiepark 25, 15232 Frankfurt (Oder), Germany

[7] Now with Institute of Photonic Systems, Swiss Federal Institute of Technology (ETH), Zurich, Switzerland

We report on the first experimental demonstration of a GaAs modulator suited for complex IQ modulation formats. Data rates of up to 150 Gbit/s are generated using QPSK, 16-QAM, 32-QAM and 64-QAM on a single carrier and polarization.

5.2.1 Introduction

The deployment of advanced modulation formats in next-generation optical transport network systems (OTN) has been the key driver for increasing spectral efficiency and scaling network capacity [154]. In this context, service providers are currently upgrading or transforming existing interfaces to scalable and efficient digital optical platforms featuring high-bandwidth photonic integrated circuits (PIC) [155]. The path to channel rates of 100 Gbit/s and beyond involves modules that employ high-speed IQ modulators that feature high integration density, low driving voltages and low-cost fabrication.

So far $LiNbO_3$ IQ modulators have been widely exploited in high-speed communication modules due to the strong electro-optic effect and high speed operation. Data rates of more than 100 Gbit/s were achieved with quadrature phase-shift keying (QPSK) and various quadrature amplitude modulation (QAM) formats [156]. However, while footprint is one of the key parameters and newer, smaller modules are currently being considered and standardized, $LiNbO_3$ devices exhibit a relatively large chip size. As an alternative, compact InP modulators can be used, enabling, e.g., dual-polarization-(DP-)QPSK transmission at symbol rates of 28 GBd [157]. However, although InP has the potential to meet the form factor requirements, cost and yield pose challenges for commercialization, as fabrication relies on 2-inch InP wafers. GaAs, in contrast, is a mature technology widely used for monolithic integration of microwave integrated circuits on large-area 6-inch wafers. The material system offers high electron mobility, low-loss semi-insulating substrates, operation and cost-efficient foundry-based fabrication. In addition, GaAs lends itself to integration of photonic devices such as passive waveguides, photo detectors and lasers. GaAs addresses the demands of next-generation OTN interfaces. Nevertheless, while GaAs technology is also suited for monolithic integration of low loss high-speed Mach-Zehnder electro-optic modulators [158], IQ modulators have not yet been shown.

In this paper we report on the first experimental demonstration of an IQ modulator exploiting the linear electro-optic effect in GaAs. Using 64-QAM at a symbol rate of 25 GBd we achieve a data rate of 150 Gbit/s. This is to the best of our knowledge the first time that GaAs modulators are used for coherent communication with higher-order modulation formats resulting in the highest data rate demonstrated with such devices so far. We further evaluate the performance of the device for a variety of application-relevant modulation formats such as QPSK, 16-QAM, 32-QAM, and 64-QAM.

5.2.2 GaAs Modulator

The GaAs modulators were produced using u2t Photonics' foundry-based fabri-fabrication processes. Six-inch semi-insulating GaAs wafers are used in the processing of the modulators. The epi-layers are grown by molecular beam epitaxy (MBE). Standard i-line steppers are used for photolithography to define optical waveguides and electrodes. The basic structure of the IQ modulator is depicted in Fig. 5.3a. It consists of two high-speed MZI modulators, taking equal optical inputs from a splitter, and outputting to a recombiner via phase-shift electrodes. The configuration may be considered as an outer 'parent' interferometer comprising two inner 'child' MZI units. The two RF drives are independent, ideally, with no crosstalk. The IQ modulator includes optimized deep-ridge to shallow-rib transitions, shallow-rib S-bends, deep-ridge corners and U-bends, and many other features designed for low loss and high extinction ratio. The design is tolerant to process variation and maintains its performance over a wide range of geometric parameters. The MZIs are based on Y-branch optical splitters and combiners to achieve high extinction ratio (ER) and broadband operation resulting in devices with improved temperature stability, reliable sinusoidal transfer functions and highly reproducible device performance [158]. The tested IQ chip comprises the child and parent phase-control electrodes, recombiners and monitor taps.

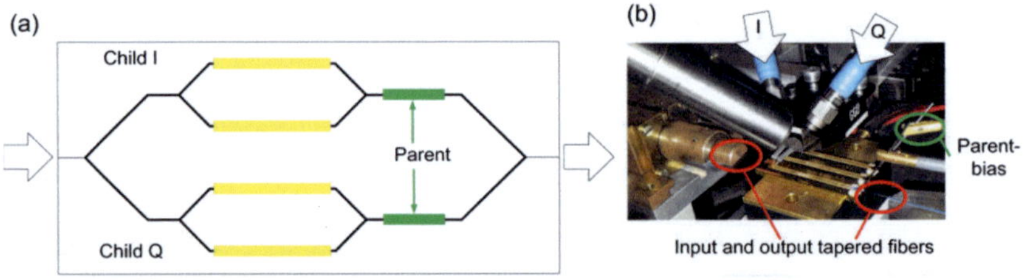

Fig. 5.3: Fabricated GaAs IQ modulator. (a) Schematic layout of two child Mach-Zehnder modulators (in yellow), which deliver the in-phase (I) and quadrature-phase (Q) modulation, respectively, and are nested in a parent Mach-Zehnder interferometer (MZI). Lensed polarization maintaining fibers were used to couple the light to and from the chip. Operating point adjustments of the child MZMs are done by adding a bias voltage to the RF probes. Dedicated phase shifters are used to set the angle of I and Q modulation in the parent MZI. (b) Picture of the device under test in characterization setup.

With 30 mm long RF electrodes, V_π is below 3V; nevertheless, the modulation bandwidth is typically larger than 30 GHz owing to the low-loss coplanar waveguide (CPW) configuration and excellent velocity matching. To demonstrate the raw performance of the chip, single IQ modulator chips were mounted on a brass carrier and the CPW lines were terminated with a bonded off-chip resistor.

5.2.3 Experimental Setup

The experimental setup is depicted in Fig. 5.4: The electrical driving signal was generated by two Virtex-5 FPGA-boards connected each to a Micram DAC25, functioning as an arbitrary waveform generator (AWG) [145]. The analog electrical signal was amplified by two SHF electrical amplifiers to a peak-to-peak voltage of 2.5 V_{pp}. This signal was coupled to the CPW by a RF probe. The operating points of the inner MZIs were set by applying a bias to the RF electrodes. The substrate-voltage and parent-bias were applied by tungsten needles.

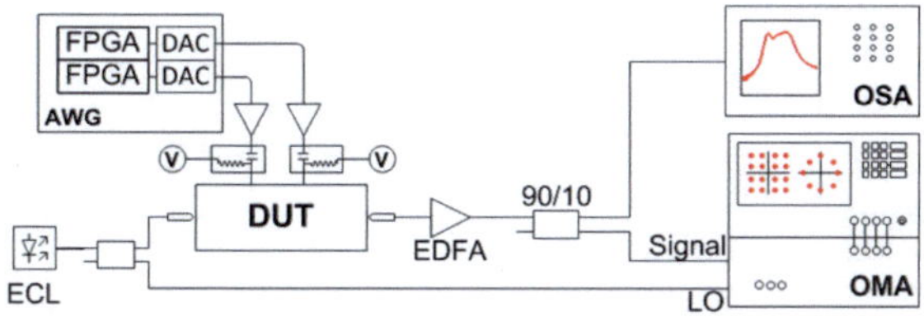

Fig. 5.4: Experimental Setup: Two Virtex5-FPGA-boards each linked to a Micram DAC25 are used as an arbitrary waveform generator providing the signal. The outputs for the I- and Q-arm are amplified before being coupled to the chip by a picoprobe. An ECL serves as optical source and is coupled to the chip by a tapered fiber. The modulated signal is amplified by an EDFA and received by an OMA. Light of the original ECL is used as LO to avoid frequency offset. An OSA was attached to a 10 dB-coupler right after the out-coupling tapered fiber.

Light was coupled to and from the device by using polarization-maintaining lensed fibers. The device is operated at a wavelength of 1539.4 nm and an input power of 7 dBm. The insertion loss is better than 18 dB, including coupling losses inherent to the test setup. The modulator itself has an insertion loss of approximately 8 dB. After the device the optical signal was amplified by an erbium doped fiber amplifier (EDFA) and fed to an Agilent N4391A optical modulation analyzer (OMA) for coherent detection and signal quality evaluation. The transmitter laser was used as a local oscillator (LO) in the OMA.

5.2.4 Experimental Results

The experimental results are depicted in Fig. 5.5. The back-to-back characterization was performed at the best OSNR we could achieve (> 25 dB) and using equalization with 55 taps at the receiver (1 tap per symbol). For 32-QAM and 64-QAM an 8-tap (1 tap per symbol) pre-emphasis was employed as well. Constellation diagrams of QPSK (a), 16-QAM (b), 32-QAM (c) and 64-QAM (d) are depicted with a histogram of 51200 points per plot. The error vector magnitude (EVM, maximum normalization) was determined to be 6.1 %, 7.1 %, 6.3 % and 6.9 %, respectively. Error-free QPSK generation is proven, bit error-ratios for 16-QAM and 32-QAM are well below hard-decision threshold limits for forward-error correction (FEC). The measured EVM for 64-QAM translates to a BER which indicates that data transmission and reception using a software-decision threshold should be possible.

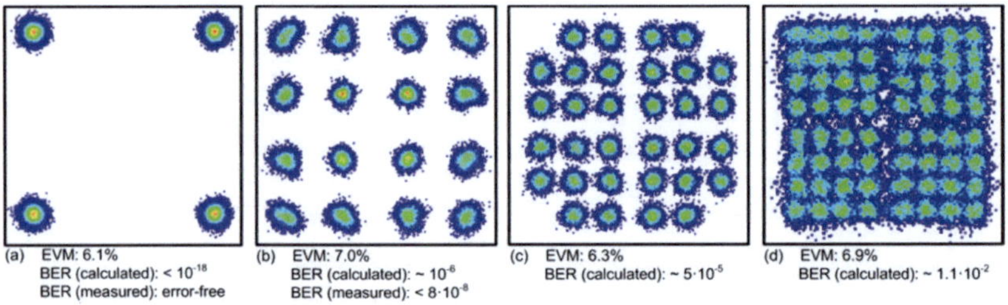

Fig. 5.5: Constellation diagrams generated with GaAs IQ modulator. Bit-error ratios (BER) have been measured and in part calculated from EVM [146], [147]. Subfigures show (a) QPSK, (b) 16-QAM, (c) 32-QAM and (d) 64-QAM respectively. For QPSK the EVM is determined to be 6.1%. No bit errors were found over minutes. For 16-QAM, an EVM of 7.0% is measured and also no bit errors could be found within a recording time of 125 µs (BER < $8\cdot10^{-8}$). For 32-QAM, the EVM is 6.3%. For 64-QAM the EVM is 6.9%. Except for 64-QAM all bit-error ratios lie well below the level of 2nd generation hard-decision forward error correction (FEC) limits [159], while 64-QAM would have to rely on a soft-decision FEC.

5.2.5 Conclusion

We have experimentally demonstrated a high-speed GaAs IQ-modulator. The device was operated at a symbol rate of 25 GBd using various higher-order modulation formats such as QPSK, 16-QAM, 32-QAM and 64-QAM. This leads to data rates of up to 150 Gbit/s. To our knowledge this is the first time that complex modulation was demonstrated in a GaAs-based IQ-modulator.

[End of Paper]

5.3 Summary

Measurements have been made with:

- SOH modulators with chromophores hosted in a polymer matrix,
- SOH modulators with organic crystals,
- SOI pn-modulators, and
- GaAs modulators.

Experience and reports are available for

- $LiNbO_3$ modulators, and
- InP based modulators [3].

Combining the observations and experience from the measurements and reported modulator performance, a very rough comparison of modulators from different platforms is attempted. Note that the importance of many parameters depends strongly on the intended application. Also, Si modulators have not yet reached the maturity of their counterparts in $LiNbO_3$, InP or GaAs, which have a head start. Hence we present our perception and simply estimate the near-term potential of each technology.

Table 5-1 presents an overview of our opinion on the different platforms. Some observations are discussed in detail in the following: GaAs and $LiNbO_3$ show the highest modulation quality and may be the best choice when integration functionality, space and energy consumption are of secondary interest. InP has a business case when integration functionality is desired and on-chip lasers with high yield are required, as in the case with DWDM networks.

Hybrid integration of germanium (for detection), and III-V dies (to make lasers) is possible with SOI plasma-effect-based modulators. This is interesting for optical interconnects, i.e., in cases where volume and cost per unit count most.

SOH modulators provide good modulation quality with no nearby physical limit preventing further amelioration and excellent, low drive voltages. The use of organic materials in the SOH platform has to be investigated for reliability. Telecordia certification (a reliability certification) has been obtained for packaged polymer modulators (waveguides are completely made from polymers) using the optically nonlinear material M3 of GigOptix. The very same material has been used to make SOH modulators in this work and it is conceivable that they could show the same reliability as polymer modulators when packaged. In the short-term, the best selling point for SOH modulators is their very low π-voltage which enables modulators with an energy consumption down to 1.6 fJ/bit [C2], at 12.5 Gbit/s, which is impressive compared to pn-modulators with the current record at 200 fJ/bit at 20 Gbit/s [160]. The low π-voltage can also be exploited in frequency comb line generation, which requires to drive modulators at multiples of the π-voltage at high frequencies. By designing the SOH device with a length of 0.5 µm or less, it can be driven as a lumped element and part of the low π-voltage can be traded for higher bandwidth. This makes modulators with exceptionally large bandwidth possible [C7]. High-bandwidth and low π-voltage modulators are of interest in analog applications, like microwave photonics to enable the optical processing of electronic signals, e.g. for phased-array radar systems [161].

Parameter	**Si (organic hybrid)**	**Si (plasma effect)**	**GaAs**	**InP**	**$LiNbO_3$**	**Impact on customer value and cost of ownership**
Relative size	1 mm	3 mm	3 cm	2 mm [3]	8 cm	high
Bias stability	– drift at low freq., suspected to be a result of moving ions	O	++	+	O	low
$V_\pi L$	0.5 Vmm	10 Vmm	90 Vmm	15 Vmm	90 Vmm	high
ER (DC)	25 dB	less than 25 dB	25 dB	less than 25 dB	25 dB	high
Modulation formats	16QAM	QPSK (16QAM)	64QAM	16QAM	64QAM	medium
Integration functionality	++	++	+	++	– –	high
Linearity[17]	+ Pockels effect, low frequency issues	– Plasma effect, depletion zone variation	++ Pockels effect	O quantum-confined Stark effect (QCSE)	+ Pockels effect, low frequency issues	medium
Wavelength range	+	+	+	O wavelength dependence of QCSE	+	high
Vol. scaling	+ foundries available	+ foundries available	+ foundries available	–	– –	medium
Wafer size	8”…12”	8”…12”	6”	2”…4”	n.a.	

Table 5-1: List of estimated figures for modulator performance of different platforms.

[17] The linearity is defined as the dependence of an induced phase shift vs. applied voltage.

6 Conclusions

The demonstrations made in this work illustrate the main advantages of the silicon-organic hybrid (SOH) platform. The silicon part of the photonic integrated circuits (PIC) is made with fabrication techniques known from the well-established CMOS industry. Making organic claddings in a post-processing step allows building on this CMOS fabrication infrastructure. The organic claddings can deliver properties which are not available otherwise from silicon-on-insulator (SOI) waveguides. SOH modulators based on nonlinear chromophores hosted in polymers have been demonstrated using advanced modulation formats at 112 Gbit/s. In contrast to other platforms, which are limited to a certain set of materials, one can further develop the organic materials, and can proceed beyond mere optimization. This enables the creation of disruptive devices. For instance, by having demonstrated an SOH modulator based on organic crystals in this work, we advocate a new material class for making high-speed modulators on SOI. The organic crystals also lend themselves to applications such as THz-wave generation. The performance of SOH modulators has been briefly compared to the cases of SOI plasma-effect modulators and GaAs modulators. SOH modulators show promise for applications requiring ultra-low energy consumption, or compatibility with low-voltage driving circuits. Ordinarily employed electrical amplifiers are thus made obsolete, as demonstrated by directly driving an SOH modulator with a SiGe circuit. Generation of light has been achieved using the SOH platform by making an optically pumped on-chip SOH laser. This laser exceeds the state of the art by an order of magnitude in terms of emission pulse peak power. The SOH platform offers the prospect to make this kind of laser mass-producible.

In conclusion we find that the SOH platform is destined to improve data communication, especially when the platform's potential for parallelization will be realized. For instance, multiplexing several data sub-channels by employing OFDM could be realized in a very compact form. The required sub-components for the PIC such as comb line generators, modulators and optical Fourier transform are already all available. Beyond this domain, applications in the fields of metrology, life-sciences and microwave photonics could profit from the SOH platform.

Appendix A: Supplementary Information

A.1 Useful Relations

The following relation is known for a vector **E**, which depends on the spatial coordinate **r**:

$$\begin{aligned}\nabla^2\mathcal{E}(\mathbf{r}) &= \nabla\left(\nabla\cdot\mathcal{E}(\mathbf{r})\right)-\nabla\times\nabla\times\mathcal{E}(\mathbf{r}) \\ &= \mathrm{grad}\left(\mathrm{div}\left(\mathcal{E}(\mathbf{r})\right)\right)-\mathrm{curl}\left(\mathrm{curl}\left(\mathcal{E}(\mathbf{r})\right)\right)\end{aligned} \tag{7.1}$$

The following relation for small x with $x, a \in \mathbb{R}$ is useful:

$$\frac{1}{\sqrt{a^2+x}} \simeq \frac{1}{a}-\frac{x}{2a^3}. \tag{7.2}$$

A useful identity for $a, b \in \mathbb{C}$ is

$$\frac{1}{2}\left(\exp(\mathrm{j}a)+\exp(\mathrm{j}b)\right)=\exp\left(\frac{\mathrm{j}a+\mathrm{j}b}{2}\right)\cos\left(\frac{\mathrm{j}a-\mathrm{j}b}{2}\right). \tag{7.3}$$

A.2 Definitions

The Fourier transform used in this thesis is defined as

$$\begin{aligned}\tilde{\mathbf{E}}(\mathbf{r},\omega) &= \mathrm{F}\{\mathbf{E}(\mathbf{r},t)\} = \int_{\mathbb{R}}\mathbf{E}(\mathbf{r},t)\exp(-\mathrm{j}\omega t)\mathrm{d}t\ , \\ \mathbf{E}(\mathbf{r},t) &= \mathrm{F}^{-1}\{\tilde{\mathbf{E}}(\mathbf{r},\omega)\} = \frac{1}{2\pi}\int_{\mathbb{R}}\tilde{\mathbf{E}}(\mathbf{r},\omega)\exp(\mathrm{j}\omega t)\mathrm{d}\omega\ .\end{aligned} \tag{7.4}$$

First and second moments can be calculated for a density function $f(x)$. The first moment is

$$\overline{x} = \int_{\mathbb{R}} x f(x)\,\mathrm{d}x. \tag{7.5}$$

The second central moment is

$$\sigma_{\mathrm{2nd}}^2 = \int_{\mathbb{R}}\left(x-\overline{x}\right)^2 f(x)\,\mathrm{d}x. \tag{7.5}$$

The complementary error function is

$$\mathrm{erfc}(z)=\frac{2}{\sqrt{\pi}}\int_z^{\infty}\exp(-t^2)\,\mathrm{d}t. \tag{7.5}$$

The Gaussian beam has the normalized *field* distribution $f_{\mathrm{E}}(x)$

$$\int_{\mathbb{R}} f_{\mathrm{E}}(x)\,\mathrm{d}x = \int_{\mathbb{R}} \frac{1}{\sqrt{2\pi\sigma_{\mathrm{2nd,E}}^2}} \exp\left(\frac{-x^2}{2\sigma_{\mathrm{2nd,E}}^2}\right)\mathrm{d}x = 1. \tag{7.6}$$

Its beam radius r is defined at the $1/e$ fraction of the peak field value as $r=\sqrt{2}\sigma_{\mathrm{2nd,E}}$, and hence the diameter is $D_{1/e,\mathrm{E}} = 2r = 2\sqrt{2}\sigma_{\mathrm{2nd,E}}$.

The corresponding normalized *intensity* distribution $f_{\mathrm{I}}(x)$ can be written

$$\int_{\mathbb{R}} f_{\mathrm{I}}(x)\,\mathrm{d}x = \int_{\mathbb{R}} \frac{1}{\sqrt{2\pi\sigma_{\mathrm{2nd,I}}^2}} \exp\left(\frac{-x^2}{2\sigma_{\mathrm{2nd,I}}^2}\right)\mathrm{d}x = 1. \tag{7.7}$$

The standard variation of both distributions are then related as $\sqrt{2}\sigma_{\mathrm{2nd,I}} = \sigma_{\mathrm{2nd,E}}$ and a new diameter $D_{4\sigma}$ at the $1/e^2$ fraction of the peak intensity value can be defined as

$$D_{4\sigma} = D_{1/e^2,\mathrm{I}} = 4\sigma_{\mathrm{2nd,I}} = 2\sqrt{2}\sigma_{\mathrm{2nd,E}} = D_{1/e,\mathrm{E}} = 2r. \tag{7.8}$$

The new diameter $D_{4\sigma}$ is twice the radius of the Gaussian beam. As $\sigma_{\mathrm{2nd,I}}^2$ is also the second moment of the Gaussian *intensity* distribution, this is called its second moment width, which is identically with the usual definition of the Gaussian beam waist. This second moment width can also be calculated for a distribution which is not Gaussian using the second central moment definition of Eq. (7.5). Hence, it is often used to compare non-Gaussian beams with Gaussian beams.

Quality Metrics for Optical Signals

Besides counting errors in sufficient numbers to derive the bit-error-ratio (BER) of a communication link, other metrics can be employed to estimate the quality of the link. An overview about the most commonly used methods is presented in [42], [162].

Using a (an optical) digital communication analyzer (DCA) which samples a repetitive optical signal, one can obtain an eye diagram (a histogram of samples over the symbol duration, for an example see Fig. 4.4(b)). In case of direct detection as used for OOK, one can obtain the quality factor corresponding to Q^2 from this eye diagram, by looking at the distribution of the detector voltage into the levels of "1" and "0". The quantity Q^2 approximately equals the electrical signal-to-noise power ratio, provided that no inter-symbol interference is present, ones and zeros are uniformly distributed, the first moment of a "0" is 0, the second moments fulfill the conditions $1 < \sigma_{V,"1"}^2 / \sigma_{V,"0"}^2 \le 2$ and $\sigma_{V,"1"}^2 < \left(\sigma_{V,"1"}^2 + \sigma_{V,"0"}^2\right) / 2 \le 1.5\sigma_{V,"1"}^2$. The quantity Q^2 is then close to the ratio of the difference of the first moments of the detector voltage ($\overline{V}_{"1"}$, $\overline{V}_{"0"}$, which are proportional to the light intensity) and the sum of the second moments (square root) of the detector voltage ($\sigma_{V,"1"}, \sigma_{V,"0"}$).

$$Q^2 \simeq \left(\frac{\overline{V}_{"1"} - \overline{V}_{"0"}}{\sigma_{V,"1"} + \sigma_{V,"0"}} \right)^2 \tag{7.9}$$

It can be shown that the BER can be estimated from the quality factor, using the complementary error function (see definition above), as

$$\mathrm{BER} = \frac{1}{2} \mathrm{erfc}\left(\sqrt{\frac{Q^2}{2}} \right) \tag{7.10}$$

In case of coherent detection, the so-called error vector magnitude (EVM) can be used as a quality metric for quadratic m-ary QAM signals [146], [147]. The underlying idea is to look at the ith received complex amplitude $\mathbf{E}_{\mathrm{r},\,i}$ and compare it with the originally transmitted amplitude $\mathbf{E}_{\mathrm{t},\,i}$ in a constellation diagram, in order to obtain the deviation $\mathbf{E}_{\mathrm{err},\,i}$ from a perfect transmission:

$$\mathbf{E}_{\mathrm{err},\,i} = \mathbf{E}_{\mathrm{r},\,i} - \mathbf{E}_{\mathrm{t},\,i} \tag{7.11}$$

By performing this measurement N_S times, the variance σ_{err}^2 is given, which can also be written as root mean square deviation σ_{err}. By dividing this quantity with the absolute, maximum transmitted amplitude $\left|\mathbf{E}_{\mathrm{t,\,max}}\right|$, a normalization is performed. The resulting quantity is called $\mathrm{EVM}_{\max}$ (in maximum notation)

$$\mathrm{EVM}_{\max} = \frac{\sigma_{\mathrm{err}}}{|\mathbf{E}_{\mathrm{t,\,max}}|}, \text{ with } \sigma_{\mathrm{err}}^2 = \frac{1}{N_S} \sum_{i=1}^{N_S} |\mathbf{E}_{\mathrm{err},\,i}|^2 . \tag{7.12}$$

A factor k can be defined using the averaged (over the power of the used symbols) absolute amplitude $\left|\mathbf{E}_{\mathrm{t,\,a}}\right|$ as

$$k = \frac{|\mathbf{E}_{\mathrm{t,\,m}}|}{|\mathbf{E}_{\mathrm{t,\,a}}|} . \tag{7.13}$$

For different modulation formats $E_{\mathrm{t,\,a}}$ differs, hence:

$$\begin{array}{c|c|c|c|c} & \text{QPSK} & \text{16QAM} & \text{32QAM} & \text{64QAM} \\ \hline k^2 & 1 & 9/5 & 17/10 & 7/3 \end{array}. \tag{7.14}$$

According to [146] the EVM corresponds to a certain BER, assuming additive white Gaussian noise, and non-data-aided reception.

$$\mathrm{BER} = \frac{1-m^{-1/2}}{\frac{1}{2}\log_2 m}\,\mathrm{erfc}\left(\frac{3/2}{(m-1)\left(k\,\mathrm{EVM}_{\mathrm{max}}\right)^2}\right)^{1/2} \tag{7.15}$$

Appendix B: References

[1] D. Kilper, "Energy Efficient Networks," in *Optical Fiber Communication Conference/National Fiber Optic Engineers Conference 2011*, 2011, p. OWI5.

[2] A. A. Saleh, "Evolution of the Architecture and Technology of Data Centers towards Exascale and Beyond," in *Optical Fiber Communication Conference/National Fiber Optic Engineers Conference 2013*, 2013, p. OM2D.4.

[3] R. A. Griffin, S. K. Jones, N. Whitbread, S. C. Heck, and L. N. Langley, "InP Mach-Zehnder Modulator Platform for 10/40/100/200-Gb/s Operation," *IEEE J. Sel. Top. Quantum Electron.*, vol. 19, no. 6, pp. 3401209–3401209, 2013.

[4] R. A. Griffin, A. Tipper, and I. Betty, "Performance of MQW InP Mach-Zehnder modulators for advanced modulation formats," in *Optical Fiber Communication Conference, 2005. Technical Digest. OFC/NFOEC*, 2005, vol. 2, p. 3 pp. Vol. 2–.

[5] G. T. Reed, G. Mashanovich, F. Y. Gardes, and D. J. Thomson, "Silicon optical modulators," *Nat. Photon.*, vol. 4, no. 8, pp. 518–526, Jul. 2010.

[6] J. Michel, J. Liu, and L. C. Kimerling, "High-performance Ge-on-Si photodetectors," *Nat. Photonics*, vol. 4, no. 8, pp. 527–534, Jul. 2010.

[7] J. Michel, R. E. Camacho-Aguilera, Y. Cai, N. Patel, J. T. Bessette, M. Romagnoli, R. Dutt, and L. Kimerling, "An Electrically Pumped Ge-on-Si Laser," in *National Fiber Optic Engineers Conference*, 2012, p. PDP5A.6.

[8] J. Leuthold, W. Freude, J.-M. Brosi, R. Baets, P. Dumon, I. Biaggio, M. L. Scimeca, F. Diederich, B. Frank, and C. Koos, "Silicon organic hybrid technology—a platform for practical nonlinear optics," *Proc. IEEE*, vol. 97, no. 7, pp. 1304–1316, 2009.

[9] J. F. Bauters, M. L. Davenport, M. J. R. Heck, J. K. Doylend, A. Chen, A. W. Fang, and J. E. Bowers, "Silicon on ultra-low-loss waveguide photonic integration platform," *Opt. Express*, vol. 21, no. 1, p. 544, Jan. 2013.

[10] D. Liang and J. E. Bowers, "Recent progress in lasers on silicon," *Nat Photon*, vol. 4, no. 8, pp. 511–517, 2010.

[11] B. Corcoran, C. Monat, C. Grillet, D. J. Moss, B. J. Eggleton, T. P. White, L. O'Faolain, and T. F. Krauss, "Green light emission in silicon through slow-light enhanced third-harmonic generation in photonic-crystal waveguides," *Nat. Photon.*, vol. 3, no. 4, pp. 206–210, Apr. 2009.

[12] A. Shakoor, R. Lo Savio, P. Cardile, S. L. Portalupi, D. Gerace, K. Welna, S. Boninelli, G. Franzò, F. Priolo, T. F. Krauss, M. Galli, and L. O'Faolain, "Room temperature all-silicon photonic crystal nanocavity light emitting diode at sub-bandgap wavelengths," *Laser Photonics Rev.*, vol. 7, no. 1, pp. 114–121, 2013.

[13] R. Baets, "Photonics: An ultra-small silicon laser," *Nature*, vol. 498, no. 7455, pp. 447–448, Jun. 2013.

[14] H. Rong, R. Jones, A. Liu, O. Cohen, D. Hak, A. Fang, and M. Paniccia, "A continuous-wave Raman silicon laser," *Nature*, vol. 433, no. 7027, pp. 725–728, Feb. 2005.

[15] H. Rong, S. Xu, Y.-H. Kuo, V. Sih, O. Cohen, O. Raday, and M. Paniccia, "Low-threshold continuous-wave Raman silicon laser," *Nat Photon*, vol. 1, no. 4, pp. 232–237, Apr. 2007.

[16] Y. Takahashi, Y. Inui, M. Chihara, T. Asano, R. Terawaki, and S. Noda, "A micrometre-scale Raman silicon laser with a microwatt threshold," *Nature*, vol. 498, no. 7455, pp. 470–474, Jun. 2013.

[17] J. Liu, X. Sun, R. Camacho-Aguilera, L. C. Kimerling, and J. Michel, "Ge-on-Si laser operating at room temperature," *Opt. Lett.*, vol. 35, no. 5, pp. 679–681, Mar. 2010.

[18] R. E. Camacho-Aguilera, Y. Cai, N. Patel, J. T. Bessette, M. Romagnoli, L. C. Kimerling, and J. Michel, “An electrically pumped germanium laser,” *Opt. Express*, vol. 20, no. 10, pp. 11316–11320, May 2012.

[19] H. Park, A. Fang, S. Kodama, and J. Bowers, “Hybrid silicon evanescent laser fabricated with a silicon waveguide and III-V offset quantum wells,” *Opt. Express*, vol. 13, no. 23, pp. 9460–9464, Nov. 2005.

[20] A. W. Fang, R. Jones, H. Park, O. Cohen, O. Raday, M. J. Paniccia, and J. E. Bowers, “Integrated AlGaInAs-silicon evanescent race track laser and photodetector,” *Opt. Express*, vol. 15, no. 5, pp. 2315–2322, Mar. 2007.

[21] R. Bonk, “Linear and Nonlinear Semiconductor Optical Amplifiers for Next-Generation Optical Networks,” *PhD Thesis Karlsruhe Inst. Technol. KIT*, vol. 8, 2013.

[22] D. Hillerkuss, R. Schmogrow, T. Schellinger, M. Jordan, M. Winter, G. Huber, T. Vallaitis, R. Bonk, P. Kleinow, F. Frey, M. Roeger, S. Koenig, A. Ludwig, A. Marculescu, J. Li, M. Hoh, M. Dreschmann, J. Meyer, S. B. Ezra, N. Narkiss, B. Nebendahl, F. Parmigiani, P. Petropoulos, B. Resan, A. Oehler, K. Weingarten, T. Ellermeyer, J. Lutz, M. Moeller, M. Huebner, J. Becker, C. Koos, W. Freude, and J. Leuthold, “26 Tbit s-1 line-rate super-channel transmission utilizing all-optical fast Fourier transform processing,” *Nat. Photon.*, vol. 5, no. 6, pp. 364–371, Jun. 2011.

[23] P. Spruyt and S. Vanhastel, “The Numbers are in: Vectoring 2.0 Makes G.fast Faster | TechZine | Alcatel-Lucent 07/04/2013,” *TechZine - Alcatel-Lucent*, 04-Jul-2013. [Online]. Available: http://www2.alcatel-lucent.com/techzine/the-numbers-are-in-vectoring-2-0-makes-g-fast-faster/. [Accessed: 16-Aug-2013].

[24] D. J. Thomson, F. Y. Gardes, J.-M. Fedeli, S. Zlatanovic, Y. Hu, B. P. P. Kuo, E. Myslivets, N. Alic, S. Radic, G. Z. Mashanovich, and G. T. Reed, “50-Gb/s silicon optical modulator,” *Photon. Technol. Lett. IEEE*, vol. 24, no. 4, pp. 234 –236, Feb. 2012.

[25] L. Alloatti, R. Palmer, S. Diebold, K. P. Pahl, B. Chen, R. Dinu, M. Fournier, J.-M. Fedeli, T. Zwick, W. Freude, C. Koos, and J. Leuthold, “100 GHz silicon–organic hybrid modulator,” *Light Sci. Appl.*, vol. 3, no. 5, p. e173, May 2014.

[26] P. Dong, X. Liu, C. Sethumadhavan, L. L. Buhl, R. Aroca, Y. Baeyens, and Y.-K. Chen, “224-Gb/s PDM-16-QAM Modulator and Receiver based on Silicon Photonic Integrated Circuits,” in *Optical Fiber Communication Conference/National Fiber Optic Engineers Conference 2013*, 2013, p. PDP5C.6.

[27] D. M. Pozar, *Microwave Engineering*, 4th Edition. Wiley, 2011.

[28] R. W. Boyd, *Nonlinear optics*. San Diego (California): Academic Press, 2008.

[29] T. Vallaitis, “Ultrafast Nonlinear Silicon Waveguides and Qantum Dot Semiconductor Optical Amplifiers,” *PhD Thesis Karlsruhe Inst. Technol. KIT*, vol. 7, 2011.

[30] R. A. Soref and B. R. Bennett, “Electrooptical effects in silicon,” *IEEE J. Quantum Electron.*, vol. 23, no. 1, pp. 123–129, 1987.

[31] Q. Lin, O. J. Painter, and G. P. Agrawal, “Nonlinear optical phenomena in silicon waveguides: modeling and applications,” *Opt. Express*, vol. 15, no. 25, pp. 16604–16644, Dec. 2007.

[32] R. R. Vardanyan, V. K. Dallakyan, U. Kerst, and C. Boit, “Modeling free carrier absorption in silicon,” *J. Contemp. Phys. Armen. Acad. Sci.*, vol. 47, no. 2, pp. 73–79, Apr. 2012.

[33] L. Alloatti, M. Lauermann, C. Sürgers, C. Koos, W. Freude, and J. Leuthold, “Optical absorption in silicon layers in the presence of charge inversion/accumulation or ion implantation,” *Appl. Phys. Lett.*, vol. 103, no. 5, pp. 051104–051104–5, Jul. 2013.

[34] C. Koos, P. Vorreau, T. Vallaitis, P. Dumon, W. Bogaerts, R. Baets, B. Esembeson, I. Biaggio, T. Michinobu, F. Diederich, W. Freude, and J. Leuthold, "All-optical high-speed signal processing with silicon–organic hybrid slot waveguides," *Nat. Photon.*, vol. 3, no. 4, pp. 216–219, Mar. 2009.

[35] J.-M. Brosi, C. Koos, L. C. Andreani, M. Waldow, J. Leuthold, and W. Freude, "High-speed low-voltage electro-optic modulator with a polymer-infiltrated silicon photonic crystal waveguide," *Opt. Express*, vol. 16, no. 6, pp. 4177–4191, Mar. 2008.

[36] C. Koos, "Nanophotonic Devices for Linear and Nonlinear Optical Signal Processing," *PhD Thesis Karlsruhe Inst. Technol. KIT*, 2007.

[37] Y. Vlasov and S. McNab, "Losses in single-mode silicon-on-insulator strip waveguides and bends," *Opt. Express*, vol. 12, no. 8, pp. 1622–1631, Apr. 2004.

[38] W. Bogaerts and S. K. Selvaraja, "Compact single-mode silicon hybrid rib/strip waveguide with adiabatic bends," *IEEE Photonics J.*, vol. 3, no. 3, pp. 422 –432, Jun. 2011.

[39] G. Chiarotti, S. Nannarone, R. Pastore, and P. Chiaradia, "Optical Absorption of Surface States in Ultrahigh Vacuum Cleaved (111) Surfaces of Ge and Si," *Phys. Rev. B*, vol. 4, no. 10, pp. 3398–3402, Nov. 1971.

[40] F. van Laere, T. Claes, J. Schrauwen, S. Scheerlinck, W. Bogaerts, D. Taillaert, L. O'Faolain, D. Van Thourhout, and R. Baets, "Compact focusing grating couplers for silicon-on-insulator integrated circuits," *IEEE Photonics Technol. Lett.*, vol. 19, no. 23, pp. 1919 –1921, Dec. 2007.

[41] Y. Quan, P. Han, Q. Ran, F. Zeng, L. Gao, and C. Zhao, "A photonic wire-based directional coupler based on SOI," *Opt. Commun.*, vol. 281, no. 11, pp. 3105–3110, Jun. 2008.

[42] D. Hillerkuss, "Single-Laser Multi-Terabit/s Systems," *PhD Thesis Karlsruhe Inst. Technol. KIT*, 2012.

[43] J. E. Bowers, D. Liang, A. W. Fang, H. Park, R. Jones, and M. J. Paniccia, "Hybrid Silicon Lasers: The Final Frontier to Integrated Computing," *Opt. Photonics News*, vol. 21, no. 5, pp. 28–33, May 2010.

[44] M. Lauermann, "Polymer Based Active Silicon Organic Hybrid Devices," *Master Thesis Karlsruhe Inst. Technol. KIT*, 2011.

[45] R. Schaefer and C. Willis, "Quantum-mechanical theory of the organic-dye laser," *Phys. Rev. A*, vol. 13, no. 5, pp. 1874–1890, May 1976.

[46] F. P. Schafer, *Dye lasers*. New York, NY (USA); Springer-Verlag New York Inc., 1990.

[47] F. J. Duarte and L. W. Hillman, *Dye laser principles, with applications*. San Diego, CA (United States); Academic Press Inc., 1990.

[48] T. Elsaesser and W. Kaiser, "Infrared dye lasers for the wavelength range 1–2 μm," in *Dye Lasers: 25 Years*, D. M. Stuke, Ed. Springer Berlin Heidelberg, 1992, pp. 95–109.

[49] F. P. Schäfer, "Dye lasers and laser dyes in physical chemistry," in *Dye Lasers: 25 Years*, vol. 70, M. Stuke, Ed. Berlin, Heidelberg: Springer Berlin Heidelberg, 1992, pp. 19–36.

[50] A. E. Vasdekis, S. A. Moore, A. Ruseckas, T. F. Krauss, I. D. W. Samuel, and G. A. Turnbull, "Silicon based organic semiconductor laser," *Appl. Phys. Lett.*, vol. 91, p. 051124, 2007.

[51] T. Morishita, K. Yamashita, H. Yanagi, and K. Oe, "1.3 μm Solid-State Plastic Laser in Dye-Doped Fluorinated Polyimide Waveguide," *Appl. Phys. Express*, vol. 3, p. 092202, 2010.

[52] W. Kranitzky, B. Kopainsky, W. Kaiser, K. H. Drexhage, and G. A. Reynolds, "A new infrared laser dye of superior photostability tunable to 1.24 μm with picosecond excitation," *Opt. Commun.*, vol. 36, no. 2, pp. 149–152, Jan. 1981.

[53] Exciton, "IR26 - Manufacturer information." [Online]. Available: http://www.exciton.com/pdfs/IR26-0511.pdf. [Accessed: 02-Sep-2013].

[54] J.-M. Brosi, "Slow-Light Photonic Crystal Devices for High-Speed Optical Signal Processing," *PhD Thesis Karlsruhe Inst. Technol. KIT*, vol. 4, 2009.

[55] L. R. Dalton, A. W. Harper, B. Wu, R. Ghosn, J. Laquindanum, Z. Liang, A. Hubbel, and C. Xu, "Polymeric Electro-Optic Modulators: Matereials synthesis and processing," *Adv. Mater.*, vol. 7, no. 6, pp. 519–540, 1995.

[56] L. R. Dalton, P. A. Sullivan, and D. H. Bale, "Electric Field Poled Organic Electro-optic Materials: State of the Art and Future Prospects," *Chem. Rev.*, vol. 110, no. 1, pp. 25–55, Jan. 2010.

[57] J. Witzens, T. Baehr-Jones, and M. Hochberg, "Design of transmission line driven slot waveguide Mach-Zehnder interferometers and application to analog optical links," *Opt. Express*, vol. 18, no. 16, p. 16902, Jul. 2010.

[58] H. Yu and W. Bogaerts, "An Equivalent Circuit Model of the Traveling Wave Electrode for Carrier-Depletion-Based Silicon Optical Modulators," *J. Light. Technol.*, vol. 30, no. 11, pp. 1602–1609, 2012.

[59] I.-L. Bundalo, "Optimisation of Electro-optic Silicon-Organic Hybrid Modulators for High-Bandwidth Applications," *Master Thesis Karlsruhe Inst. Technol. KIT*, 2012.

[60] L. Alloatti, "Silicon Photonics for High-Speed, Low-Power and Mid-IR Applications," *PhD Thesis Karlsruhe Inst. Technol. KIT*, vol. 12, 2013.

[61] O. Boyraz and B. Jalali, "Demonstration of a silicon Raman laser," *Opt Express*, vol. 12, no. 21, pp. 5269–5273, 2004.

[62] B. Jalali and S. Fathpour, "Silicon Photonics," *Light. Technol. J. Of*, vol. 24, no. 12, pp. 4600–4615, Dec. 2006.

[63] M. Hochberg and T. Baehr-Jones, "Towards fabless silicon photonics," *Nat. Photon.*, vol. 4, no. 8, pp. 492–494, 2010.

[64] H. Park, A. W. Fang, O. Cohen, R. Jones, M. J. Paniccia, and J. E. Bowers, "A Hybrid AlGaInAs-Silicon Evanescent Amplifier," *IEEE Photonics Technol. Lett.*, vol. 19, no. 4, pp. 230 –232, Feb. 2007.

[65] A. W. Fang, H. Park, O. Cohen, R. Jones, M. J. Paniccia, and J. E. Bowers, "Electrically pumped hybrid AlGaInAs-silicon evanescent laser," *Opt. Express*, vol. 14, no. 20, pp. 9203–9210, Oct. 2006.

[66] D. Liang, A. Fang, H.-W. Chen, M. Sysak, B. Koch, E. Lively, O. Raday, Y.-H. Kuo, R. Jones, and J. Bowers, "Hybrid silicon evanescent approach to optical interconnects," *Appl. Phys. Mater. Sci. Process.*, vol. 95, no. 4, pp. 1045–1057, 2009.

[67] S. M. Chen, M. C. Tang, J. Wu, Q. Jiang, V. G. Dorogan, M. Benamara, Y. I. Mazur, G. J. Salamo, A. J. Seeds, and H. Liu, "1.3 um InAs/GaAs quantum-dot laser monolithically grown on Si substrates operating over 100 °C;," *Electron. Lett.*, vol. 50, no. 20, pp. 1467–1468, Sep. 2014.

[68] S. Wirths, R. Geiger, N. von den Driesch, G. Mussler, T. Stoica, S. Mantl, Z. Ikonic, M. Luysberg, S. Chiussi, J. M. Hartmann, H. Sigg, J. Faist, D. Buca, and D. Grützmacher, "Lasing in direct-bandgap GeSn alloy grown on Si," *Nat. Photon.*, vol. 9, no. 2, pp. 88–92, Feb. 2015.

[69] C. A. Barrios and M. Lipson, "Electrically driven silicon resonant light emitting device based on slot-waveguide," *Opt. Express*, vol. 13, no. 25, pp. 10092–10101, Dec. 2005.

[70] P. Pintus, S. Faralli, and F. D. Pasquale, "Low-Threshold Pump Power and High Integration in Al O:Er Slot Waveguide Lasers on SOI," *IEEE Photonics Technol. Lett.*, vol. 22, no. 19, pp. 1428–1430, Oct. 2010.

[71] A. Tengattini, D. Gandolfi, N. Prtljaga, A. Anopchenko, J. M. Ramirez, F. F. Lupi, Y. Berencen, D. Navarro-Urrios, P. Rivallin, K. Surana, B. Garrido, J. Fedeli, and L. Pavesi, "Toward a 1.54 m Electrically Driven Erbium-Doped Silicon Slot Waveguide and Optical Amplifier," *J. Light. Technol.*, vol. 31, no. 3, pp. 391–397, Feb. 2013.

[72] H. Isshiki, F. Jing, T. Sato, T. Nakajima, and T. Kimura, "Rare earth silicates as gain media for silicon photonics [Invited]," *Photonics Res.*, vol. 2, no. 3, p. A45, Jun. 2014.

[73] E. S. Hosseini, Purnawirman, J. D. B. Bradley, J. Sun, G. Leake, T. N. Adam, D. D. Coolbaugh, and M. R. Watts, "CMOS-compatible 75 mW erbium-doped distributed feedback laser," *Opt. Lett.*, vol. 39, no. 11, p. 3106, Jun. 2014.

[74] X. Sun, A. Zadok, M. J. Shearn, K. A. Diest, A. Ghaffari, H. A. Atwater, A. Scherer, and A. Yariv, "Electrically pumped hybrid evanescentSi/InGaAsP lasers," *Opt. Lett.*, vol. 34, no. 9, p. 1345, May 2009.

[75] S. Tanaka, S.-H. Jeong, S. Sekiguchi, T. Kurahashi, Y. Tanaka, and K. Morito, "High-output-power, single-wavelength silicon hybrid laser using precise flip-chip bonding technology," *Opt. Express*, vol. 20, no. 27, p. 28057, Dec. 2012.

[76] M. Lauermann, R. Palmer, S. Koeber, P. C. Schindler, D. Korn, T. Wahlbrink, J. Bolten, M. Waldow, D. L. Elder, L. R. Dalton, J. Leuthold, W. Freude, and C. Koos, "Low-power silicon-organic hybrid (SOH) modulators for advanced modulation formats," *Opt. Express*, vol. 22, no. 24, p. 29927, Dec. 2014.

[77] R. Palmer, S. Koeber, D. L. Elder, M. Woessner, W. Heni, D. Korn, M. Lauermann, W. Bogaerts, L. Dalton, W. Freude, J. Leuthold, and C. Koos, "High-Speed, Low Drive-Voltage Silicon-Organic Hybrid Modulator Based on a Binary-Chromophore Electro-Optic Material," *J. Light. Technol.*, vol. 32, no. 16, pp. 2726–2734, Aug. 2014.

[78] J. Pfeifle, L. Alloatti, W. Freude, J. Leuthold, and C. Koos, "Silicon-organic hybrid phase shifter based on a slot waveguide with a liquid-crystal cladding," *Opt. Express*, vol. 20, no. 14, pp. 15359–15376, Jul. 2012.

[79] X. Wang, S. Grist, J. Flueckiger, N. A. F. Jaeger, and L. Chrostowski, "Silicon photonic slot waveguide Bragg gratings and resonators," *Opt. Express*, vol. 21, no. 16, p. 19029, Aug. 2013.

[80] Q. Xu, V. R. Almeida, R. R. Panepucci, and M. Lipson, "Experimental demonstration of guiding and confining light in nanometer-sizelow-refractive-index material," *Opt. Lett.*, vol. 29, no. 14, pp. 1626–1628, Jul. 2004.

[81] S. K. Selvaraja, W. Bogaerts, P. Dumon, D. Van Thourhout, and R. Baets, "Subnanometer Linewidth Uniformity in Silicon Nanophotonic Waveguide Devices Using CMOS Fabrication Technology," *IEEE J. Sel. Top. Quantum Electron.*, vol. 16, no. 1, pp. 316 –324, Feb. 2010.

[82] I. D. W. Samuel, E. B. Namdas, and G. A. Turnbull, "How to recognize lasing," *Nat Photon*, vol. 3, no. 10, pp. 546–549, Oct. 2009.

[83] T. Elsaesser and W. Kaiser, "Infrared dye lasers for the wavelength range 1–2 μm," in *Dye Lasers: 25 Years*, vol. 70, M. Stuke, Ed. Berlin, Heidelberg: Springer Berlin Heidelberg, 1992, pp. 95–109.

[84] J. Zhang and Z. Zhu, "Novel heptamethine thiapyrylium infrared laser dyes of superior photostability tunable from 1.35 to 1.65 μm," *Opt. Commun.*, vol. 113, no. 1–3, pp. 61–64, Dec. 1994.

[85] Y. Zhang and S. R. Forrest, "Existence of continuous-wave threshold for organic semiconductor lasers," *Phys. Rev. B*, vol. 84, no. 24, p. 241301, Dec. 2011.

[86] H. Schmidt and A. R. Hawkins, "The photonic integration of non-solid media using optofluidics," *Nat. Photon.*, vol. 5, no. 10, pp. 598–604, 2011.

[87] R. D. Schaller, M. A. Petruska, and V. I. Klimov, "Tunable Near-Infrared Optical Gain and Amplified Spontaneous Emission Using PbSe Nanocrystals," *J. Phys. Chem. B*, vol. 107, no. 50, pp. 13765–13768, Dec. 2003.

[88] A. L. Rogach, A. Eychmüller, S. G. Hickey, and S. V. Kershaw, "Infrared-Emitting Colloidal Nanocrystals: Synthesis, Assembly, Spectroscopy, and Applications," *Small*, vol. 3, no. 4, pp. 536–557, 2007.

[89] J. Heo, Z. Jiang, J. Xu, and P. Bhattacharya, "Coherent and directional emission at 1.55 μm from PbSe colloidal quantum dot electroluminescent device on silicon," *Opt. Express*, vol. 19, no. 27, p. 26394, Dec. 2011.

[90] M. Iqbal, M. A. Gleeson, B. Spaugh, F. Tybor, W. G. Gunn, M. Hochberg, T. Baehr-Jones, R. C. Bailey, and L. C. Gunn, "Label-Free Biosensor Arrays Based on Silicon Ring Resonators and High-Speed Optical Scanning Instrumentation," *Ieee J. Sel. Top. Quantum Electron.*, vol. 16, no. 3, pp. 654–661, Jun. 2010.

[91] M. Casalboni, F. De Matteis, P. Prosposito, A. Quatela, and F. Sarcinelli, "Fluorescence efficiency of four infrared polymethine dyes," *Chem. Phys. Lett.*, vol. 373, no. 3–4, pp. 372–378, May 2003.

[92] D. P. Benfey, D. C. Brown, S. J. Davis, L. G. Piper, and R. F. Foutter, "Diode-pumped dye laser analysis and design," *Appl. Opt.*, vol. 31, no. 33, pp. 7034–7041, Nov. 1992.

[93] "CST - Computer Simulation Technology. 3d electromagnetic simulation software." [Online]. Available: https://www.cst.com/. [Accessed: 10-May-2015].

[94] T. Vallaitis, S. Bogatscher, L. Alloatti, P. Dumon, R. Baets, M. L. Scimeca, I. Biaggio, F. Diederich, C. Koos, W. Freude, and J. Leuthold, "Optical properties of highly nonlinear silicon-organic hybrid (SOH) waveguide geometries," *Opt. Express*, vol. 17, no. 20, pp. 17357–17368, 2009.

[95] O. E. Semonin, J. C. Johnson, J. M. Luther, A. G. Midgett, A. J. Nozik, and M. C. Beard, "Absolute Photoluminescence Quantum Yields of IR-26 Dye, PbS, and PbSe Quantum Dots," *J. Phys. Chem. Lett.*, vol. 1, no. 16, pp. 2445–2450, Aug. 2010.

[96] C. V. Shank, "Physics of dye lasers," *Rev. Mod. Phys.*, vol. 47, no. 3, pp. 649–657, Jul. 1975.

[97] C. Koos, L. Jacome, C. Poulton, J. Leuthold, and W. Freude, "Nonlinear silicon-on-insulator waveguides for all-optical signal processing," *Opt. Express*, vol. 15, no. 10, pp. 5976–5990, May 2007.

[98] J.-M. Brosi, C. Koos, L. C. Andreani, M. Waldow, J. Leuthold, and W. Freude, "High-speed low-voltage electro-optic modulator with a polymer-infiltrated silicon photonic crystal waveguide," *Opt. Express*, vol. 16, no. 6, pp. 4177–4191, Mar. 2008.

[99] C. Koos, L. Jacome, C. Poulton, J. Leuthold, and W. Freude, "Nonlinear silicon-on-insulator waveguides for all-optical signal processing," *Opt. Express*, vol. 15, no. 10, p. 5976, 2007.

[100] B. W. Hakki and T. L. Paoli, "Gain spectra in GaAs double−heterostructure injection lasers," *J. Appl. Phys.*, vol. 46, no. 3, p. 1299, 1975.

[101] D. P. Benfey, D. C. Brown, S. J. Davis, L. G. Piper, and R. F. Foutter, "Diode-pumped dye laser analysis and design," *Appl. Opt.*, vol. 31, no. 33, p. 7034, Nov. 1992.

[102] A. C. Turner-Foster, M. A. Foster, J. S. Levy, C. B. Poitras, R. Salem, A. L. Gaeta, and M. Lipson, "Ultrashort free-carrier lifetime in low-loss silicon nanowaveguides," *Opt. Express*, vol. 18, no. 4, p. 3582, Feb. 2010.

[103] M. Dinu, F. Quochi, and H. Garcia, "Third-order nonlinearities in silicon at telecom wavelengths," *Appl. Phys. Lett.*, vol. 82, no. 18, pp. 2954–2956, May 2003.

[104] T. Vallaitis, S. Bogatscher, L. Alloatti, P. Dumon, R. Baets, M. L. Scimeca, I. Biaggio, F. Diederich, C. Koos, W. Freude, and J. Leuthold, "Optical properties of highly nonlinear silicon-organic hybrid (SOH) waveguide geometries," *Opt. Express*, vol. 17, no. 20, pp. 17357–17368, 2009.

[105] Y. Koizumi, K. Toyoda, M. Yoshida, and M. Nakazawa, "1024 QAM (60 Gbit/s) single-carrier coherent optical transmission over 150 km," *Opt. Express*, vol. 20, no. 11, p. 12508, May 2012.

[106] J. Leuthold and W. Freude, "Chapter 9 - Optical OFDM and Nyquist Multiplexing," in *Optical Fiber Telecommunications (Sixth Edition)*, I. P. Kaminow, T. Li, and A. E. Willner, Eds. Boston: Academic Press, 2013, pp. 381–432.

[107] W. M. Green, M. J. Rooks, L. Sekaric, and Y. A. Vlasov, "Ultra-compact, low RF power, 10 Gb/s siliconMach-Zehnder modulator," *Opt. Express*, vol. 15, no. 25, pp. 17106–17113, Dec. 2007.

[108] L. Liao, A. Liu, D. Rubin, J. Basak, Y. Chetrit, H. Nguyen, R. Cohen, N. Izhaky, and M. Paniccia, "40 Gbit/s silicon optical modulator for high-speed applications," *Electron. Lett.*, vol. 43, no. 22, pp. 1196–1197, Oct. 2007.

[109] O.-P. Kwon, S.-J. Kwon, M. Jazbinsek, F. D. J. Brunner, J.-I. Seo, C. Hunziker, A. Schneider, H. Yun, Y.-S. Lee, and P. Günter, "Organic phenolic configurationally locked polyene single crystals for electro-optic and terahertz wave applications," *Adv. Funct. Mater.*, vol. 18, no. 20, pp. 3242–3250, 2008.

[110] R. S. Jacobsen, K. N. Andersen, P. I. Borel, J. Fage-Pedersen, L. H. Frandsen, O. Hansen, M. Kristensen, A. V. Lavrinenko, G. Moulin, H. Ou, C. Peucheret, B. Zsigri, and A. Bjarklev, "Strained silicon as a new electro-optic material," *Nature*, vol. 441, no. 7090, pp. 199–202, May 2006.

[111] B. Chmielak, M. Waldow, C. Matheisen, C. Ripperda, J. Bolten, T. Wahlbrink, M. Nagel, F. Merget, and H. Kurz, "Pockels effect based fully integrated, strained silicon electro-optic modulator," *Opt. Express*, vol. 19, no. 18, p. 17212, Aug. 2011.

[112] C. Matheisen, M. Nagel, S. Sawallich, M. Waldow, B. Chmielak, T. Wahlbrink, J. Bolten, and H. Kurz, "Locally induced electro-optic activity in silicon nanophotonic devices," *CLEOEUROPE - IQEC*, 2013.

[113] J. H. Wülbern, A. Petrov, and M. Eich, "Electro-optical modulator in a polymerinfiltrated silicon slotted photonic crystal waveguide heterostructure resonator," *Opt. Express*, vol. 17, no. 1, pp. 304–313, Jan. 2009.

[114] H. Chen, B. Chen, D. Huang, D. Jin, J. D. Luo, A. K.-Y. Jen, and R. Dinu, "Broadband electro-optic polymer modulators with high electro-optic activity and low poling induced optical loss," *Appl. Phys. Lett.*, vol. 93, no. 4, pp. 043507–043507–3, Jul. 2008.

[115] Rainbow Photonics, "Rainbow Photonics AG - Terahertz Sources, Terahertz Spectroscopy, Terahertz Imaging, Organic Electro-Optic Crystals, Tunable Infrared Lasers." [Online]. Available: http://www.rainbowphotonics.com/. [Accessed: 02-Sep-2013].

[116] C. Hunziker, S.-J. Kwon, H. Figi, M. Jazbinsek, and P. Günter, "Fabrication and phase modulation in organic single-crystallineconfigurationally locked, phenolic polyene OH1 waveguides," *Opt. Express*, vol. 16, no. 20, pp. 15903–15914, Sep. 2008.

[117] H. Figi, D. H. Bale, A. Szep, L. R. Dalton, and A. Chen, "Electro-optic modulation in horizontally slotted silicon/organic crystal hybrid devices," *J. Opt. Soc. Am. B*, vol. 28, no. 9, pp. 2291–2300, Sep. 2011.

[118] H. Figi, M. Jazbinsek, C. Hunziker, M. Koechlin, and P. Günter, "Electro-optic tuning and modulation of single-crystalline organic microring resonators," *J. Opt. Soc. Am. B*, vol. 26, no. 5, pp. 1103–1110, May 2009.

[119] M. Fujiwara, M. Maruyama, M. Sugisaki, H. Takahashi, S. Aoshima, R. J. Cogdell, and H. Hashimoto, "Determination of the d-tensor components of a single crystal of N-benzyl-2-methyl-4-nitroaniline," *Jpn. J. Appl. Phys.*, vol. 46, no. 4A, pp. 1528–1530, 2007.

[120] J. Witzens, T. Baehr-Jones, and M. Hochberg, "Design of transmission line driven slot waveguide Mach-Zehnder interferometers and application to analog optical links," *Opt. Express*, vol. 18, no. 16, p. 16902, Jul. 2010.

[121] R. Ding, T. Baehr-Jones, Y. Liu, R. Bojko, J. Witzens, S. Huang, J. Luo, S. Benight, P. Sullivan, J.-M. Fedeli, M. Fournier, L. Dalton, A. Jen, and M. Hochberg, "Demonstration of a low V_πL modulator with GHz bandwidth based on electro-optic polymer-clad silicon slot waveguides," *Opt. Express*, vol. 18, no. 15, pp. 15618–15623, Jul. 2010.

[122] D. Rezzonico, S.-J. Kwon, H. Figi, O.-P. Kwon, M. Jazbinsek, and P. Günter, "Photochemical stability of nonlinear optical chromophores in polymeric and crystalline materials," *J. Chem. Phys.*, vol. 128, no. 12, p. 124713, Mar. 2008.

[123] M. Jazbinsek, H. Figi, C. Hunziker, B. Ruiz, S.-J. Kwon, O.-P. Kwon, Z. Yang, and P. Günter, "Organic electro-optic single crystalline films for integrated optics," presented at the Linear and Nonlinear Optics of Organic Materials X, San Diego, California, USA, 2010, p. 77740Q–77740Q–10.

[124] P. Günter, *Nonlinear Optical Effects and Materials*, vol. 72. Springer, 2000.

[125] M. Jazbinsek, L. Mutter, and P. Gunter, "Photonic applications with the organic nonlinear optical crystal DAST," *IEEE J. Sel. Top. Quantum Electron.*, vol. 14, no. 5, pp. 1298–1311, 2008.

[126] L. Mutter, F. D. Brunner, Z. Yang, M. Jazbinsek, and P. Günter, "Linear and nonlinear optical properties of the organic crystal DSTMS," *J. Opt. Soc. Am. B*, vol. 24, no. 9, pp. 2556–2561, Sep. 2007.

[127] A. Chen and E. J. Murphy, *Broadband Optical Modulators: Science, Technology, and Applications*. CRC Press, 2011.

[128] H. Figi, M. Jazbinsek, C. Hunziker, M. Koechlin, and P. Günter, "Electro-optic single-crystalline organic waveguides and nanowires grown from the melt," *Opt. Express*, vol. 16, no. 15, pp. 11310–11327, Jul. 2008.

[129] D. Vermeulen, S. Selvaraja, P. Verheyen, G. Lepage, W. Bogaerts, P. Absil, D. Van Thourhout, and G. Roelkens, "High-efficiency fiber-to-chip grating couplers realized using an advanced CMOS-compatible Silicon-On-Insulator platform," *Opt. Express*, vol. 18, no. 17, pp. 18278–18283, Aug. 2010.

[130] M. Pu, L. Liu, H. Ou, K. Yvind, and J. M. Hvam, "Ultra-low-loss inverted taper coupler for silicon-on-insulator ridge waveguide," *Opt. Commun.*, vol. 283, no. 19, pp. 3678–3682, Oct. 2010.

[131] W. M. Green, M. J. Rooks, L. Sekaric, and Y. A. Vlasov, "Ultra-compact, low RF power, 10 Gb/s siliconMach-Zehnder modulator," *Opt. Express*, vol. 15, no. 25, pp. 17106–17113, Dec. 2007.

[132] A. Brimont, D. J. Thomson, P. Sanchis, J. Herrera, F. Y. Gardes, J. M. Fedeli, G. T. Reed, and J. Martí, "High speed silicon electro-optical modulators enhanced via slow light propagation," *Opt. Express*, vol. 19, no. 21, pp. 20876–20885, Oct. 2011.

[133] J. H. Wülbern, J. Hampe, A. Petrov, M. Eich, J. Luo, A. K.-Y. Jen, A. Di Falco, T. F. Krauss, and J. Bruns, "Electro-optic modulation in slotted resonant photonic crystal heterostructures," *Appl. Phys. Lett.*, vol. 94, no. 24, p. 241107, 2009.

[134] K. Padmaraju, N. Ophir, Q. Xu, B. Schmidt, J. Shakya, S. Manipatruni, M. Lipson, and K. Bergman, "Error-free transmission of microring-modulated BPSK," *Opt. Express*, vol. 20, no. 8, pp. 8681–8688, Apr. 2012.

[135] P. Dong, C. Xie, L. Chen, N. K. Fontaine, and Y. Chen, "Experimental demonstration of microring quadrature phase-shift keying modulators," *Opt. Lett.*, vol. 37, no. 7, pp. 1178–1180, Apr. 2012.

[136] P. Dong, C. Xie, L. Chen, L. L. Buhl, and Y.-K. Chen, "112-Gb/s monolithic PDM-QPSK modulator in silicon," *Opt. Express*, vol. 20, no. 26, p. B624, Dec. 2012.

[137] R. S. Jacobsen, K. N. Andersen, P. I. Borel, J. Fage-Pedersen, L. H. Frandsen, O. Hansen, M. Kristensen, A. V. Lavrinenko, G. Moulin, H. Ou, C. Peucheret, B. Zsigri, and A. Bjarklev, "Strained silicon as a new electro-optic material," *Nature*, vol. 441, no. 7090, pp. 199–202, May 2006.

[138] T. Baehr-Jones, M. Hochberg, G. Wang, R. Lawson, Y. Liao, P. Sullivan, L. Dalton, A. Jen, and A. Scherer, "Optical modulation and detection in slotted Silicon waveguides," *Opt. Express*, vol. 13, no. 14, pp. 5216–5226, Jul. 2005.

[139] J. H. Wülbern, A. Petrov, and M. Eich, "Electro-optical modulator in a polymer-infiltrated silicon slotted photonic crystal waveguide heterostructure resonator," *Opt. Express*, vol. 17, no. 1, pp. 304–313, Jan. 2009.

[140] I. A. Young, E. M. Mohammed, J. T. S. Liao, A. M. Kern, S. Palermo, B. A. Block, M. R. Reshotko, and P. L. D. Chang, "Optical technology for energy efficient I/O in high performance computing," *IEEE Commun. Mag.*, vol. 48, no. 10, pp. 184–191, 2010.

[141] P. Dong, L. Chen, and Y. Chen, "High-speed low-voltage single-drive push-pull silicon Mach-Zehnder modulators," *Opt. Express*, vol. 20, no. 6, p. 6163, Feb. 2012.

[142] A. Brimont, D. J. Thomson, F. Y. Gardes, J. M. Fedeli, G. T. Reed, J. Martí, and P. Sanchis, "High-contrast 40Gb/s operation of a 500 µm long silicon carrier-depletion slow wave modulator," *Opt. Lett.*, vol. 37, no. 17, p. 3504, Aug. 2012.

[143] D. Jin, H. Chen, A. Barklund, J. Mallari, G. Yu, E. Miller, and R. Dinu, "EO polymer modulators reliability study," p. 75990H–75990H, Feb. 2010.

[144] R. Blum, M. Sprave, J. Sablotny, and M. Eich, "High-electric-field poling of nonlinear optical polymers," *J. Opt. Soc. Am. B*, vol. 15, no. 1, pp. 318–328, Jan. 1998.

[145] R. Schmogrow, D. Hillerkuss, M. Dreschmann, M. Huebner, M. Winter, J. Meyer, B. Nebendahl, C. Koos, J. Becker, W. Freude, and J. Leuthold, "Real-Time Software-Defined Multiformat Transmitter Generating 64QAM at 28 GBd," *IEEE Photonics Technol. Lett.*, vol. 22, no. 21, pp. 1601 –1603, Nov. 2010.

[146] R. Schmogrow, B. Nebendahl, M. Winter, A. Josten, D. Hillerkuss, S. Koenig, J. Meyer, M. Dreschmann, M. Huebner, C. Koos, J. Becker, W. Freude, and J. Leuthold, "Error Vector Magnitude as a Performance Measure for Advanced Modulation Formats," *IEEE Photonics Technol. Lett.*, vol. 24, no. 1, pp. 61 –63, Jan. 2012.

[147] R. Schmogrow, B. Nebendahl, M. Winter, A. Josten, D. Hillerkuss, S. Koenig, J. Meyer, M. Dreschmann, M. Huebner, C. Koos, J. Becker, W. Freude, and J. Leuthold, "Corrections ibid.: Error Vector Magnitude as a Performance Measure for Advanced Modulation Formats," *IEEE Photonics Technol. Lett.*, vol. 24, no. 23, pp. 2198–2198, 2012.

[148] J. Proakis and M. Salehi, *Digital Communications*. McGraw-Hill Education, 2007.

[149] L. Dalton and S. Benight, "Theory-Guided Design of Organic Electro-Optic Materials and Devices," *Polymers*, vol. 3, no. 4, pp. 1325–1351, Aug. 2011.

[150] J. K. Doylend, P. E. Jessop, and A. P. Knights, "Silicon photonic resonator-enhanced defect-mediated photodiode for sub-bandgap detection," *Opt. Express*, vol. 18, no. 14, pp. 14671–14678, Jul. 2010.

[151] M. W. Geis, S. J. Spector, M. E. Grein, J. U. Yoon, D. M. Lennon, and T. M. Lyszczarz, "Silicon waveguide infrared photodiodes with >35 GHz bandwidth and phototransistors with 50 AW-1 response," *Opt. Express*, vol. 17, no. 7, pp. 5193–5204, Mar. 2009.

[152] D. F. Logan, P. E. Jessop, and A. P. Knights, "Modeling Defect Enhanced Detection at 1550 nm in Integrated Silicon Waveguide Photodetectors," *J. Light. Technol.*, vol. 27, no. 7, pp. 930–937, Apr. 2009.

[153] Y. Zhao, C. Xu, W.-J. Wang, Q. Zhou, Y.-L. Hao, J.-Y. Yang, M.-H. Wang, and X.-Q. Jiang, "Photocurrent Effect in Reverse-Biased p-n Silicon Waveguides in Communication Bands," *Chin. Phys. Lett.*, vol. 28, p. 074216, Jul. 2011.

[154] P. J. Winzer, A. H. Gnauck, C. R. Doerr, M. Magarini, and L. L. Buhl, "Spectrally Efficient Long-Haul Optical Networking Using 112-Gb/s Polarization-Multiplexed 16-QAM," *J. Light. Technol.*, vol. 28, no. 4, pp. 547–556, Feb. 2010.

[155] A. Deore, O. Turkcu, S. Ahuja, S. J. Hand, and S. Melle, "Total cost of ownership of WDM and switching architectures for next-generation 100Gb/s networks," *IEEE Commun. Mag.*, vol. 50, no. 11, pp. 179–187, 2012.

[156] H. Yamazaki, T. Yamada, T. Goh, and A. Kaneko, "PDM-QPSK Modulator With a Hybrid Configuration of Silica PLCs and LiNbO$_{3}$ Phase Modulators," *J. Light. Technol.*, vol. 29, no. 5, pp. 721–727, Mar. 2011.

[157] E. Yamada, S. Kanazawa, A. Ohki, K. Watanabe, Y. Nasu, N. Kikuchi, Y. Shibata, R. Iga, and H. Ishii, "112-Gb/s InP DP-QPSK Modulator Integrated with a Silica-PLC Polarization Multiplexing Circuit," in *Optical Fiber Communication Conference*, 2012, p. PDP5A.9.

[158] R. Walker and J. Heaton, "Gallium Arsenide Modulator Technology," in *Broadband Optical Modulators*, CRC Press, 2011, pp. 207–222.

[159] ITU, "G.975.1 : Forward error correction for high bit-rate DWDM submarine systems." [Online]. Available: http://www.itu.int/rec/T-REC-G.975.1-200402-I/en. [Accessed: 02-Sep-2013].

[160] T. Baehr-Jones, R. Ding, Y. Liu, A. Ayazi, T. Pinguet, N. C. Harris, M. Streshinsky, P. Lee, Y. Zhang, A. E.-J. Lim, T.-Y. Liow, S. H.-G. Teo, G.-Q. Lo, and M. Hochberg, "Ultralow drive voltage silicon traveling-wave modulator," *Opt. Express*, vol. 20, no. 11, p. 12014, May 2012.

[161] J. Capmany and D. Novak, "Microwave photonics combines two worlds," *Nat Photon*, vol. 1, no. 6, pp. 319–330, Jun. 2007.

[162] W. Freude, R. Schmogrow, B. Nebendahl, M. Winter, A. Josten, D. Hillerkuss, S. Koenig, J. Meyer, M. Dreschmann, M. Huebner, C. Koos, J. Becker, and J. Leuthold, "Quality metrics for optical signals: Eye diagram, Q-factor, OSNR, EVM and BER," in *2012 14th International Conference on Transparent Optical Networks (ICTON)*, 2012, pp. 1–4.

Appendix C: Glossary

C.1 List of Symbols

Constants

$c = 3.0 \times 10^{8}$ m/s	Velocity of light in vacuum
$e = 1.6 \times 10^{-19}$ C	Elementary charge
$\varepsilon_0 = 8.9 \times 10^{-12}$ F/m	Permittivity of vacuum
$\mu_0 = 12.6 \times 10^{-7}$ H/m	Permeability of vacuum
$\hbar = 1.1 \times 10^{-34}$ Js	Reduced Planck constant
$h = 6.6 \times 10^{-34}$ Js	Planck constant

Greek Symbols

$\alpha_{\text{amp}}, \alpha_{\text{RF,amp}}$	Attenuation coefficient (amplitude notation, real part of complex propagation constant)
α_{meas}	Attenuation coefficient (power), as measured
α_{Soref}	Attenuation coefficient (power) according to Soref
$\alpha_{\text{n, Vard}}$	Attenuation coefficient (power) according to Vardanyan for n-doped silicon
$\alpha_{\text{p, Vard}}$	Attenuation coefficient (power) according to Vardanyan for p-doped silicon
α_{P}	Attenuation coefficient (power)
β	(Common) propagation constant (imaginary part of complex propagation constant)
δ_{skin}	Skin depth
Γ	Interaction factor
Γ_{load}	Amplitude reflection factor for reflection between device and load
Γ_{gen}	Amplitude reflection factor for reflection between generator and unloaded (unterminated) device

$\Gamma_{\mathrm{term.dev}}$	Amplitude reflection factor for reflection between generator and terminated device
γ	Complex propagation constant
ε_r	Material dependent dielectric permittivity
$\boldsymbol{\varepsilon}_r$	Permittivity matrix
η_{QE}	Quantum efficiency of a photodiode
$\boldsymbol{\eta}$	Impermeability matrix
λ	Optical wavelength
λ_{c}	Center wavelength
σ_{RMS}	Effective (RMS, root mean square) roughness of a surface
σ	Conductivity
σ_a	Absorption cross section
$\rho(\mathbf{r},t)$	Electric charge density
$\mathbf{\Phi}$	Transfer matrix of a phase shifter
$\varphi, \varphi_m, \varphi_n$	Phase shift in a Mach-Zehnder modulator, delay interferometer
τ	Time constant
$\varphi_{\mathrm{I}}, \varphi_{\mathrm{Q}}$	Phase difference in the in-phase (I) or quadrature phase (Q) Mach-Zehnder modulator of an IQ-modulator
ω	Angular frequency
ω_{c}	Angular central frequency
ω_{RF}	Angular frequency of RF wave
$\chi^{(2)}$	Optical nonlinearity of second order
$\chi^{(3)}$	Optical nonlinearity of third order

Latin Symbols

$\mathbf{a}=(a_1,\ldots,a_4)^{\mathrm{T}}$	Normalized wave amplitude, such that $\lvert\mathbf{a}\rvert^2$ represents the incoming optical power
a	Crystallographic axis a
a_{Wi}	Constant collecting frequency independent terms in travelling wave model of Witzens et al.
$A(z,t)$	Amplitude factor of optical wave
A_{eff}	Effective mode cross section
A	Amplitude of an optical carrier or of a signal
$\mathbf{b}=(b_1,\ldots,b_4)^{\mathrm{T}}$	Normalized wave amplitude, such that $\lvert\mathbf{b}\rvert^2$ represents the outgoing optical power
b	Crystallographic axis b
BER	Bit error probability or bit error ratio
$\boldsymbol{\mathcal{B}}(\mathbf{r},t)$	Magnetic flux density
C	Capacitance
$\mathbf{c}=(c_1, c_2)^{\mathrm{T}}$	Normalized wave amplitude, such that $\lvert\mathbf{c}\rvert^2$ represents the incoming electrical power
c	Crystallographic axis c
$\mathbf{d}=(d_1,d_2)^{\mathrm{T}}$	Normalized wave amplitude, such that $\lvert\mathbf{d}\rvert^2$ represents the outgoing electrical power
$\boldsymbol{\mathcal{D}}(\mathbf{r},t)$	Electric flux density
$\mathbf{D}$	Transfer matrix for a delay
$\boldsymbol{\mathcal{E}}(\mathbf{r},t)$	Electric field vector
$\mathbf{E}_{\mathrm{t}}(x,y,\omega_{\mathrm{c}})$	Electric field distribution in the transverse (to the propagation direction) plane
$E_{\mathrm{norm}}(t)$	Normalized electric field of optical carrier from complex amplitude modulation
f_{RC}	3 dB bandwidth corresponding to an RC constant
EVM	Error vector magnitude

$\mathrm{EVM}_{\mathrm{max}}$	EVM normalized to the maximum electric field in the constellation
f	Frequency
G	Conductance
$\tilde{h}(\omega)$	Frequency response
$\boldsymbol{\mathcal{H}}(\mathbf{r},t)$	Magnetic field vector
$\mathbf{H}_{\mathrm{t}}(x,y,\omega_{\mathrm{c}})$	Magnetic field distribution in the transverse (to the propagation direction) plane
$I_{\mathrm{light}}(l)$	Intensity of light after propagation of length l in medium
$I_{0,\,\mathrm{light}}$	Intensity of light before propagation in medium
$I(t)$	Intensity of light
I	In-phase component of a signal
i_{PD}	Generated electrical current in a photo detector
$\mathcal{J}(\mathbf{r},t)$	Electric current density
l,L	Device length (unit m)
l_{eff}	Effective device length
L	Inductance (unit Henry, H) no general relation to length L
$\mathbf{M}_1$, $\mathbf{M}_2$	Transfer matrices of 2-by-2 MMIs (3 dB splitters/combiner)
$\mathbf{n}$	Normal vector (normal to a surface, context dependent)
n	Refractive index
n_{eff}	Effective refractive index
n_{g}	Group refractive index
N_{e}	Electron density
N_{h}	Hole density
N	Density of molecules

OSNR	Optical signal-to-noise ratio with fixed reference bandwidth of 0.1 nm. Signal power related to the unpolarized ASE power for this fixed reference bandwidth
$\mathcal{P}(\mathbf{r},t)$	Polarization density (in short polarization)
$\mathbf{P}_t = (P_x, P_y)$	Transverse polarization
P_{out}	Optical output power
p_t	Duty cycle
Q	Quadrature phase component of a signal
Q^2	Quality factor squared
$\mathbf{r}$	Spatial coordinate
r_{ij}	Electro-optic coefficient, i,j in {1,2,3} or in case of this work {x,y,z}
R_D	Photo detector responsivity
R	Resistance
$\mathbf{S}_{OO}$	Optical-to-optical scattering matrix
$\mathbf{S}_{EE}$	Electrical-to-electrical scattering matrix
$\mathbf{S}_{EO}$	Electrical-to-optical (electro-optic) scattering matrix
$S_{21,EO}(\omega_{RF})$	Element of matrix $\mathbf{S}_{EO}$ in row 2, column 1; this is the EO transfer function
$S_{21,V}(f)$	Transfer function contribution of the traveling wave electrodes to the overall electro-optic frequency response $S_{21,EO}(f)$ of an SOH modulator
$\mathbf{S}_{OE}$	Optical-to-electrical scattering matrix
S_i	Energy state of a molecule: i^{th} singlet state
T	Temperature
T_i	Energy state of a molecule: i^{th} triplet state
t	Timo
$\Delta\tau$	Time delay
V	Voltage

$V_{S,eff}$	Effective voltage $V_{S,eff}$ at source
V_{π}	π-voltage
V_{Gate}	Gate voltage (applied between SOI wafer substrate and optical Si WGs)
V_{pp}	AC voltage measured peak-to-peak
w	Width
Z_0	Characteristic impedance
Z_{gen}	Generator impedance
Z_{load}	Load impedance
Z_N	Wave impedance of the medium

C.2 Acronyms

16QAM	16ary quadrature amplitude modulation
64QAM	64ary quadrature amplitude modulation
ADC	Analog-to-digital converter
ALD	Atomic layer deposition
ASE	Amplified spontaneous emissions
ASK	Amplitude shift keying
AWG	Arrayed waveguide grating
AWGN	Additive white Gaussian noise
BER	Bit error ratio
BERT	Bit error tester
BNA	N-benzyl-2-methyl-4-nitroaniline, organic material, which can be grown as a crystal
BOX	Buried oxide (oxide layer in SOI wafer)
BPSK	Binary phase shift keying
BW	Bandwidth
CL	Cylindrical lens
CMOS	Complementary metal oxide semiconductor
CPW	Coplanar waveguide
cw	Continuous wave
DAC	Digital-to-analog converter
DCA	Digital communications analyzer - Agilent Oscilloscope
DI	Delay interferometer
DPSK	Differential phase shift keying
DQPSK	Differential quadrature phase shift keying
DSP	Digital signal processing
DWDM	Dense wavelength division multiplexing
ECL	External cavity laser

EDFA	Erbium doped fiber amplifier
EM	Electromagnetic
EO	Electro-optic
ER	Extinction ratio
EVM	Error vector magnitude
FCA	Free-carrier absorption
FEC	Forward error correction
FOM	Figure of merit
FPGA	Field programmable gate array
FSR	Free spectral range of a filter
FTTH	Fiber-to-the-home
FWHM	Full-width half-maximum
GC	Grating coupler
GSG	Ground-signal-ground
IC	Integrated circuits
IL	Insertion loss
InP	Indium phosphide
IQ modulator	Modulator for complex modulation
IR26	Organic dye
ISC	Inter system crossing
ISI	Intersymbol interference
ITU	International Telecommunication Union
IR	Infrared
LO	Local oscillator
M1, M3	Optically nonlinear chromophores in polymer matrix, produced by GigOptix
MAN	Metropolitan-area network

MBE	Molecular beam epitaxy
MFD	Mode field diameter
MLL	Mode-locked laser
MMI	Multi-mode interference coupler
MMIC	Monolithic microwave integrated circuits
MOSFET	Metal-oxide-semiconductor field effect transistor
MUX	Multiplexer
MZI	Mach-Zehnder interferometer
MZM	Mach-Zehnder modulator
NRZ	Non-return to zero
OFDM	Orthogonal frequency division multiplexing
OMA	Agilent optical modulation analyzer N4391A
OOK	On-off-keying
OTN	Optical transport network
OSA	Optical spectrum analyzer
OSNR	Optical signal to noise ratio
PBS	Polarization beam splitter
PD	Photo detector
PIC	Photonic integrated circuits
PM	Phase modulator
PRBS	Pseudo random bit sequence
PSK	Phase shift keying
PVA	Polyvinyl alcohol
QAM	Quadrature amplitude modulation
QPSK	Quadrature phase shift keying
RBW	Resolution bandwidth
RC	Resistance-capacitance

RF	Radio frequency
Rx	Receiver
SNR	Signal to noise ratio
SMF	Single mode fiber
SOI	Silicon-on-insulator
SOH	Silicon-organic hybrid
SOS	Silicon-on-sapphire
SPA	Single-photon absorption
SSMF	Standard single mode fiber
TE	Transverse electric
TL	Transmission line
TM	Transverse magnetic
TPA	Two-photon absorption
TW	Travelling wave
Tx	Transmitter
RMS	Root mean square
WAN	Wide area network
WDM	Wavelength division multiplexing
WG	Waveguide

Acknowledgements

This thesis is the result of my work at the Institute of Photonics and Quantum Electronics (IPQ) at the Karlsruhe Institute of Technology (KIT) and many cooperations, also with external partners from industry and academia.

A number of institutions contributed the essential resources and training: I acknowledge the support by the Karlsruhe School of Optics and Photonics (KSOP), by the DFG Center for Functional Nanostructures (CFN), by the European Commission for funding the EU-FP7 project SOFI (grant 248609) and the EU Network of Excellence euro-FOS, by ePIXfab (silicon photonics platform), by the BMBF joint project MISTRAL, by the Deutsche Forschungsgemeinschaft and Open Access Publishing Fund of Karlsruhe Institute of Technology. I am further grateful for support by the Karlsruhe Nano-Micro Facility (KNMF) and by the Light Technology Institute (KIT-LTI).

I would like to express my particular gratitude to my main supervisor, Prof. Juerg Leuthold, who guided this research. Starting with the big picture and with bold ideas he oriented my efforts to obtain the industrially most relevant results, including the targeting of SOH modulators for advanced modulation. He offered unwavering interest for all of the work's aspects and provided subsequent scientific and technical steering, while still encouraging following up my own ideas. Thanks to his organization, the IPQ also had the resources to indulge in pursuing the most interesting topics, which meant in my case plenty of fabrication runs at AMO GmbH and IMEC to design PICs. Moreover, he actively promoted the development of skills outside the lab, such as project management. By advancing a good measure of trust he included me in the coordination of a European consortium for the EC-funded project SOFI. This allowed me to rapidly gain experience in interaction with international partners and provided me with a close contact to industry. I learned to seek and value the win-win solution and the meaning of dependability in business relationships.

I am especially grateful to Prof. Wolfgang Freude, who continuously broadened my view on scientific and technical questions, starting from my physicist's point of view towards an engineer's way of finding a solution. From his clear way of formulating I learned to appreciate the effect that well-chosen words to formulate a question often help much to see a solution. I thank him for offering a virtually unlimited amount of time in time of need, in particular the countless hours spent on improving the quality of our publications. His counsel and support for approaching potential employment opportunities after this PhD thesis were very welcome, as was his support when I took up the project management of the SOFI project.

I would like to particularly thank Prof. Christian Koos, for directing the all-in investigation with full commitment into light sources on silicon and judgment on the feasibility of experiments, which led to the final characterization of GaAs modulators at just the right moment. His accuracy for identifying the best approach to tackle a problem in the lab was essential, as well as his non-exhaustible interest for discussing new ideas and right away their concrete implementation. His way of generating new and exciting concepts focusing on applications follows a clear vision, which made it easy to follow him. The enthusiasm he transmits along with these ideas motivates to go further in each particular research endeavor. I am grateful for his flexibility and attention to the balance of my research work and my non-scientific task of project management.

By granting access to their facilities, several people contributed to my work. In particular, I thank Prof. Uli Lemmer who granted access to his cleanroom and I thank his PhD students and personnel who operated a number of machines for my benefit and helped much with their experience. I thank Prof. Gernot Goll for organizing and granting access to the CFN cleanroom.

I especially thank Prof. Wim Bogaerts from the University of Ghent for all the help while making photonic integrated circuit (PIC) designs, and his development of the concept to open the CMOS metal stack in order to access the Si waveguides. His relentless support of the SOFI project at IMEC enabled many shared publications. I am very grateful to Hui Yu from the University of Ghent for the efforts he put into the fabrication and customization of the silicon PICs I used, as well as for the joint measurements on pn-modulators and the investigation of their capabilities as detectors. I also thank Pieter Dumon, Amit Khanna at IMEC, who both organized the chip fabrication runs and invested much time to help me in providing a functional, design-rule-error-free mask layout. I thank Shankar Kumar Sevaraja, for fabricating some of the chips, as well as Guy Lepage, Marianna Pantouvaki, Johan M.D. Wouters, Peter Verheyen, Joris Van Campenhout, Myriam Moelants, and Philippe Absil at IMEC. In addition, I express my gratitude to Katarzyna Komorowska and Prof. Roel Baets from the University of Ghent.

My special gratitude is directed at Mojca Jazbinsek from Rainbow Photonics, who invested much time and creativity to grow organic crystals on chips, which led to a joint publication and the success of the SOFI project. I also thank Prof. Peter Guenter from ETH in Zurich for his contribution to this accomplishment.

I like to express my gratitude to Silvan Wehrli, Jörg Wieland, who provided me with many insights and also with a SiGe driver circuit. I also thank Martin Bossard, Giulio Cangini, Raluca Dinu, Baoquan Chen from GigOptix, who provided coordination and the most essential nonlinear material.

I very much like to thank Panos Zakynthinosa and Ioannis Thomkos from Athens Information Technology Center, Athens, Greece for the countless discussions about energy consumption, the use and potential of the SOH platform, and the insights into needs and demands of systems for telecommunication, as well as their distinct contribution to the success of the SOFI project.

I thank the complete SOFI project team, which includes (in addition to the people mentioned above) Alberto Secchi and Massimiliano Dispenza from SELEX ES, and Steve Madden from CUDOS.

I am particularly indebted to u^{2}t Photonics UK, i.e., to M. F. O'Keefe, for access to and insights about the GaAs IQ modulator, N. Cameron, for helping with the preparation of the OFC postdeadline talk, and Y. Zhou, R. G. Walker, for their part in creating this marvelous device. In addition, I thank C. Stamatiadis, Prof. K. Petermann from the Joint Lab Silicon Photonics HFT4, at TUB in Berlin, Germany, as well as L. Stampoulidis, E. Kehayas, from Constelex Technology Enablers, Corallia Microelectronics Innovation Center, Athens, Greece for their contributions and L. Zimmermann from IHP in Frankfurt (Oder), Germany, for the coordination of the GaAs IQ modulator characterization.

I am very grateful to Tapani Alasaarela and Seppo Honkanen, from Helsinki University of Technology, Finland, for the cooperation we had on reducing the propagation losses in Si waveguides. These thanks extend also to A. Säynätjoki, A. Tervonen.

I am grateful having had the chance to work with and learn from the team of Hercules Avramopoulos from NTUA, namely Maria Spyropoulou, Christos Kouloumentas and Alexandros Maziotis.

I thank Luca Alloatti for countless discussions concerning research and many other matters, for his modulator groundwork, for broadcasting his enthusiasm about science and numerous technical solutions.

I thank Robert Palmer, for our cooperation to create the world's best MZM and IQ modulators on silicon, which includes his modeling of the modulator electrodes, low-loss optical transitions and naturally his share in characterization efforts.

I thank Philipp Schindler for, among many other things, the very productive and most efficient generation of results on the GaAs IQ modulator.

I want to express my gratitude to Argishti Melikan. I thank Rene Schmogorow, Swen König and David Hillerkuss for the infrastructure and help in the systems lab, and also Rene Bonk addionally for project management advice from a pragmatic angle. I thank Sebastian Köber for material related advice, Claudius Weimann for the experience we gained in comb line generation with SOH modulators, Jörg Pfeifle, Sascha Mühlbrandt, Jingshi Li, Nicole Lindenmann, Alexandra Ludwig, Aleksandar Nesic, Kira Köhnle, Djorn Karnick, and in short all my colleagues for their support and collaboration. This includes my former colleagues Thomas Vallaitis, whom I am thanking for advice on 3rd order nonlinear materials, Sean O'Duill and Arvind Mishra for his contribution to the SOFI project. I also like to thank the PhD students from other institutes, who helped by offering advice for fabricating my PICs, most namely Klaus Huska from LTI.

I particularly want to thank my former student and later colleague Matthias Lauermann for the very successful shared work on SOH lasers and his unwavering support in general and particular in puplication matters.

I especially thank my former student Moritz Baier for the shared work on the fastest SOH modulator to date, on the first high-speed organic crystal modulator, and his strong commitment.

I thank Wolfgang Heni and my former students Patrick Appel, Duc Nuygen, Ivan-Lazar Bundalo, Christoph Mattern, Matthias Jäger, for the shared exploration of the very diverse questions in silicon photonics, but also for their contribution in characterization efforts for the silicon photonics team.

I thank Mr. Bürger, Mr. Hirsch and Mr. Höhne for the often indispensable mechanical devices they build in the workshop of the IPQ. I thank Oswald Speck, Simone Dehm, Martin Winkeler and Florian Rupp for their marvelous technical support. I thank our secretary Ms. Olbrich for her work. I particularly thank Ms. Riemensperger for the great work on KIT's SOFI budget. I want to express my special gratitude to Bernadette Lehmann, for all the support and hints during my work at the IPQ, but also the nice reception at the very beginning.

List of Publications

Journal contributions

[J1] D. **Korn**, M. Lauermann, P. Appel, S. Koeber, L. Alloatti, R. Palmer, W. Freude, J. Leuthold, C. Koos, "Lasing in Silicon-Organic Hybrid Waveguides," *submitted.*

[J2] D. **Korn**, M. Jazbinsek, R. Palmer, M. Baier, L. Alloatti, H. Yu, W. Bogaerts, G. Lepage, P. Verheyen, P. Absil, P. Guenter, C. Koos, W. Freude, and J. Leuthold, "Electro-Optic Organic Crystal Silicon High-Speed Modulator," *IEEE Photonics J.*, vol. 6, no. 2, pp. 1–9, Apr. 2014.

[J3] P. C. Schindler, D. **Korn**, C. Stamatiadis, M. F. OKeefe, L. Stampoulidis, R. Schmogrow, P. Zakynthinos, R. Palmer, N. Cameron, Y. Zhou, R. G. Walker, E. Kehayas, S. Ben-Ezra, I. Tomkos, L. Zimmermann, K. Petermann, W. Freude, C. Koos, and J. Leuthold, "Monolithic GaAs Electro-Optic IQ Modulator Demonstrated at 150 Gbit/s With 64QAM," *J. Lightw. Technol.*, vol. 32, no. 4, pp. 760–765, Feb. 2014.

[J4] D. **Korn**, R. Palmer, H. Yu, P. C. Schindler, L. Alloatti, M. Baier, R. Schmogrow, W. Bogaerts, S. K. Selvaraja, G. Lepage, M. Pantouvaki, J. M. D. Wouters, P. Verheyen, J. Van Campenhout, B. Chen, R. Baets, P. Absil, R. Dinu, C. Koos, W. Freude, and J. Leuthold, "Silicon-organic hybrid (SOH) IQ modulator using the linear electro-optic effect for transmitting 16QAM at 112 Gbit/s," *Opt. Express*, vol. 21, no. 11, pp. 13219–13227, Jun. 2013.

[J5] R. Palmer, L. Alloatti, D. **Korn**, P. C. Schindler, M. Baier, J. Bolten, T. Wahlbrink, M. Waldow, R. Dinu, W. Freude, C. Koos, and J. Leuthold, "Low power Mach-Zehnder modulator in silicon-organic hybrid technology," *IEEE Photonics Technol. Lett.*, vol. 25, no. 13, pp. 1226–1229, 2013.

[J6] J. Leuthold, C. Koos, W. Freude, L. Alloatti, R. Palmer, D. Korn, J. Pfeifle, M. Lauermann, R. Dinu, S. Wehrli, M. Jazbinsek, P. Gunter, M. Waldow, T. Wahlbrink, J. Bolten, H. Kurz, M. Fournier, J.-M. Fedeli, H. Yu, and W. Bogaerts, "Silicon-Organic Hybrid Electro-Optical Devices," *IEEE J. Sel. Top. Quantum Electron.*, vol. 19, no. 6, pp. 3401413–3401413, 2013.

[J7] R. Palmer, L. Alloatti, D. **Korn**, P. C. Schindler, R. Schmogrow, W. Heni, S. Koenig, J. Bolten, T. Wahlbrink, M. Waldow, H. Yu, W. Bogaerts, P. Verheyen, G. Lepage, M. Pantouvaki, J. Van Campenhout, P. Absil, R. Dinu, W. Freude, C. Koos, and J. Leuthold, "Silicon-organic hybrid MZI modulator generating OOK, BPSK and 8-ASK signals for up to 84 Gbit/s," *IEEE Photonics J.*, vol. 5, no. 2, pp. 6600907–6600907, 2013.

[J8] R. Palmer, L. Alloatti, D. **Korn**, W. Heni, P. C. Schindler, J. Bolten, M. Karl, M. Waldow, T. Wahlbrink, W. Freude, C. Koos, and J. Leuthold, "Low-Loss Silicon Strip-to-Slot Mode Converters," *IEEE Photonics J.*, vol. 5, no. 1, pp. 2200409–2200409, 2013.

[J9] H. Yu, D. **Korn**, M. Pantouvaki, J. Van Campenhout, K. Komorowska, P. Verheyen, G. Lepage, P. Absil, D. Hillerkuss, L. Alloatti, J. Leuthold, R. Baets, and W. Bogaerts, "Using carrier-depletion silicon modulators for optical power monitoring," *Opt. Lett.*, vol. 37, no. 22, pp. 4681–4683, Nov. 2012.

[J10] H. Yu, M. Pantouvaki, J. Van Campenhout, D. **Korn**, K. Komorowska, P. Dumon, Y. Li, P. Verheyen, P. Absil, L. Alloatti, D. Hillerkuss, J. Leuthold, R. Baets, and W. Bogaerts, "Performance tradeoff between lateral and interdigitated doping patterns for high speed carrier-depletion based silicon modulators," *Opt. Express*, vol. 20, no. 12, pp. 12926–12938, Jun. 2012.

[J11] L. Alloatti, D. **Korn**, C. Weimann, C. Koos, W. Freude, and J. Leuthold, "Second-order nonlinear silicon-organic hybrid waveguides," *Opt. Express*, vol. 20, no. 18, pp. 20506–20515, Aug. 2012.

[J12] A. Säynätjoki, L. Karvonen, T. Alasaarela, D. **Korn**, L. Alloatti, A. Tervonen, R. Palmer, C. Koos, J. Leuthold, W. Freude, S.K. Honkanen, "Loss reduction of silicon slot waveguides with ALD-grown thin films," *Proc. SPIE*, vol. 8266, 82660D, Jan. 2012.

[J13] T. Alasaarela, D. **Korn**, L. Alloatti, A. Säynätjoki, A. Tervonen, R. Palmer, J. Leuthold, W. Freude, and S. Honkanen, "Reduced propagation loss in silicon strip and slot waveguides coated by atomic layer deposition," *Opt. Express*, vol. 19, no. 12, pp. 11529–11538, Jun. 2011.

[J14] L. Alloatti, D. **Korn**, R. Palmer, D. Hillerkuss, J. Li, A. Barklund, R. Dinu, J. Wieland, M. Fournier, J. Fedeli, H. Yu, W. Bogaerts, P. Dumon, R. Baets, C. Koos, W. Freude, and J. Leuthold, "42.7 Gbit/s electro-optic modulator in silicon technology," *Opt. Express*, vol. 19, no. 12, pp. 11841–11851, Jun. 2011.

Conference contributions

[C1] C. Weimann, S. Wolf, D. **Korn**, R. Palmer, S. Koeber, R. Schmogrow, P. C. Schindler, L. Alloatti, A. Ludwig, W. Heni, D. Bekele, D. L. Elder, H. Yu, W. Bogaerts, L. R. Dalton, W. Freude, J. Leuthold, and C. Koos, "Silicon-organic hybrid (SOH) frequency comb source for data transmission at 784 Gbit/s," *39th Eur. Conf. Exhib. Opt. Commun. ECOC*, no. Th.2.B.1, 2013.

[C2] R. Palmer, S. Koeber, W. Heni, D. L. Elder, D. **Korn**, H. Yu, L. Alloatti, S. Koenig, P. C. Schindler, W. Bogaerts, M. Pantouvaki, G. Lepage, P. Verheyen, J. Campenhout, P. Absil, R. Baets, L. R. Dalton, W. Freude, J. Leuthold, and C. Koos, "High-Speed Silicon-Organic Hybrid (SOH) Modulator with 1.6 fJ/bit and 180 pm/V In-Device Nonlinearity," presented at the ECOC 2013, London, UK, 2013, p. We.3.B.3.

[C3] Melikyan, M.; Alloatti, L.; Muslija, A.; Hillerkuss, D.; Schindler, P. C.; Li, J.; Palmer, R.; **Korn**, D.; Muehlbrandt, S.; Van Thourhout, D.; Chen, B.; Dinu, R.; Sommer, M.; Koos, C.; Kohl, M.; Freude, W.; Leuthold, J.; "Surface plasmon polariton high-speed modulator ," *Conf. on Lasers and Electro-Optics (CLEO'13),* San Jose (CA), USA, CTh5D.2, June 9–14, 2013. Postdeadline Paper

[C4] **Korn**, D.; Palmer, R.; Yu, H.; Schindler, P. C.; Alloatti, L.; Baier, M.; Schmogrow, R.; Bogaerts, W.; Selvaraja, S.; Lepage, G.; Pantouvaki, M.; Wouters, J.; Verheyen, P.; Van Campenhout, J.; Absil, P.; Baets, R.; Dinu, R.; Koos, C.; Freude, W.; Leuthold, J.; 'Silicon-organic hybrid (SOH) IQ modulator for 16QAM at 112 Gbit/s'; *Conference on Lasers and Electro-Optics Europe (CLEO-Europe/IQEC 2013)*, International Congress Centre Munich, Germany; Paper CK-9. 2 THU ; May 12–16, 2013.

[C5] D. **Korn**, P. C. Schindler, C. Stamatiadis, M. F. O'Keefe, L. Stampoulidis, R. M. Schmogrow, P. Zakynthinos, R. Palmer, N. Cameron, Y. Zhou, R. G. Walker, E. Kehayas, I. Tomkos, L. Zimmermann, K. Petermann, W. Freude, C. Koos, and J. Leuthold, "First monolithic GaAs IQ electro-optic modulator, demonstrated at 150 Gbit/s with 64-QAM," in *Optical Fiber Communication Conference/National Fiber Optic Engineers Conference 2013*, 2013, p. PDP5C.4. Postdeadline Paper

[C6] Palmer, R.; Alloatti, L.; **Korn**, D.; Schindler, P.C.; Schmogrow, R.; Baier, M.; Koenig, S.; Hillerkuss, D.; Bolten, J.; Wahlbrink, T.; Waldow, M.; Dinu, R.; Freude, W.; Koos, C. and Leuthold, J.; 'Silicon-Organic Hybrid (SOH) Modulator Generating up to 84 Gbit/s BPSK and M-ASK Signals'; *Optical Fiber Communication Conference (OFC2013),* Anaheim, CA; Novel Modulators (OW4J); pp. OW4J.6; March 17, 2013.

[C7] Alloatti, L.; **Korn**, D.; Pfeifle, J.; Palmer, R.; Koeber, S.; Baier, M.; Schmogrow, R.; Diebold, S.; Pahl, P.; Zwick, T.; Yu, H.; Bogaerts, W.; Baets, R.; Fournier, M.; Fedeli, J.; Dinu, R.; Koos, C.; Freude, W.; Leuthold, J.; 'Silicon-organic hybrid devices',*OPTO SPIE Photonics West (OPTO-SPIE'13)*, San Francisco (CA), USA, paper 8629-24; Feb. 2-7, 2013.

[C8] Leuthold, J.; Melikyan, A.; **Korn**, D.; Alloatti, L.; Palmer, R.; Koos, C.; Freude, W; 'Ultracompact CMOS-compatible Modulators' ;*Proc. Frontiers in Optics Conference, OSA*, Rochester, NY, paper FTu4A.1, Oct. 2012.

[C9] Leuthold, J.; Koos, C.; Freude, W.; Alloatti, L.; Palmer, R.; **Korn**, D.; Pfeifle, L.; Lauermann, L.; 'Silicon-Organic Hybrid - a path towards active silicon photonic devices' *Proc. EOS Annual Meeting (EOSAM 2012)*, paper 6441, Sept. 2012.

[C10] Koos., C.; Leuthold, J.; Freude, W.; Alloatti, L.; **Korn**, D.; Palmer, R.; Lauermann, M.; Lindenmann, N.; Koeber, S.; Pfeifle, J.; Schindler, P.C; Hillerkuss, D.; Schmogrow, R.; 'Silicon-Organic Hybrid Integration and Photonic Wire Bonding: Technologies for Terabit/s Interconnects' ; *Joint Symposium on Opto- and Microelectronic Devices and Circuits (SODC2012)*, Hangzhou, China, Sept. 24-27, 2012.

[C11] Leuthold, J.; Melikyan, A.; **Korn**, D.; Alloatti, L.; Palmer, R.; Koos, C.; Freude, W; 'Ultracompact CMOS-compatible Modulators', *Proc. Frontiers in Optics Conference, OSA*, Rochester, NY, paper FTu4A.1, Oct. 2012.

[C12] D. **Korn**, L. Alloatti, M. Lauermann, J. Pfeifle, R. Palmer, P. C. Schindler, W. Freude, C. Koos, J. Leuthold, H. Yu, W. Bogaerts, K. Komorowska, R. Baets, J. Van Campenhout, P. Verheyen, J. Wouters, M. Moelants, P. Absil, A. Secchi, M. Dispenza, M. Jazbinsek, P. Gunter, S. Wehrli, M. Bossard, P. Zakynthinos, and I. Tomkos, "Silicon-organic hybrid fabrication platform for integrated circuits," in *2012 14th International Conference on Transparent Optical Networks (ICTON)*, 2012.

[C13] Lauermann, M.; **Korn**, D.; Appel, P.; Alloatti, L.; Freude, W.; Leuthold, J.; Koos, C.; 'Silicon-organic hybrid (SOH) lasers at telecommunication wavelengths', *Conf. on Integrated Photonics Research, Silicon and Nano-Photonics (IPR'12)*, Cheyenne Mountain Resort, Colorado Springs, Colorado, USA, IPRSN, Paper IM3A.3, June 17-21, 2012.

[C14] Palmer, R.; Alloatti, L.; **Korn**, D.; Heni, W.; Schindler, P.; Bolten, J.; Karl, M.; Waldow, M.; Wahlbrink, T.; Freude, W.; Koos, C.; Leuthold, J.; 'Highly Efficient Strip-to-Slot Mode Converters', *CLEO 2012*, San Jose (CA), USA, Paper CM4M1, May 2012.

[C15] D. **Korn**, M. Lauermann, P. Appel, L. Alloatti, R. Palmer, W. Freude, J. Leuthold, and C. Koos, "First Silicon-Organic Hybrid Laser at Telecommunication Wavelength," in *CLEO: Science and Innovations*, 2012, p. CTu2I.1.

[C16] D. **Korn**, H. Yu, D. Hillerkuss, L. Alloatti, C. Mattern, W. Bogaerts, K. Komorowska, R. Baets, J. Van Campenhout, P. Verheyen, J. Wouters, M. Moelants, P. Absil, W. Freude, C. Koos, and J. Leuthold, "Detection or Modulation at 35 Gbit/s with a Standard CMOS-processed Optical Waveguide," in *Conference on Lasers and Electro-Optics 2012*, 2012, p. CTu1A.1.

[C17] Koos, C.; Leuthold, J.; Freude, W.; Alloatti, L.; **Korn**, D.; Lindenmann, N.; Palmer, R.; Pfeifle, J.; Hillerkuss, D.; Schmogrow, R.; 'Silicon Nanophotonics and Photonic Wire Bonding: Technologies for Terabit/s Interconnects', *6th ITG Workshop on Photonic Integration, Laser Optics Berlin*, March 20, 2012.

[C18] Yu, H.; Bogaerts, W.; Komorowska, K.; Baets, R.; **Korn**, D.; Alloatti, L.; Hillerkuss, D.; Koos, C.; Freude, W.; Leuthold, J.; Campenhout, J.; Verheyen, P.; Wouters, J.; Moelants, M.; Absil, P.; 'Doping geometries for 40G carrier-depletion-based silicon optical modulators', *Optical Fiber Communication Conference (OFC'12)*, Paper OW4F.4, Los Angeles (CA), USA, 04.–08.03.2012.

[C19] Freude, W.; Alloatti, L.; Melikyan, A.; Palmer, R.; **Korn**, D.; Lindenmann, N.; Vallaitis, T.; Hillerkuss, D.; Li, J.; Barklund, A.; Dinu, R.; Wieland, J.; Fournier, M.; Fedeli, J.; Walheim, S.; Leufke, P. M.; Ulrich, S.; Ye, J.; Vincze, P.; Hahn, H.; Yu, H.; Bogaerts, W.; Dumont, P.; Baets, R.; Breiten, B.; Diederich, F.; Beels, M. T.; Biaggio, I.; Schimmel, Th.; Koos, C.; Leuthold, J.; 'Nonlinear optics on the silicon platform', *Optical Fiber Communication Conference (OFC'12)*, Paper OTh3H.6 Los Angeles (CA), USA, 04.–08.03.2012.

[C20] Schindler, P.; Fischer, H.; Haisch, H.; Lindenmann, N.; Alloatti, L.; Palmer, R.; **Korn**, D.; Hillerkuss, D.; Schmogrow, R.; Gerster, A.; Freude, W.; Leuthold, J.; Koos, C.; 'Optical Technologies for Off-Chip and On-Chip Interconnects', *presented at DesignCon 2012*, Santa Clara (CA), USA, CA 1/31/2012, January 30 - February 1, 2012.

[C21] Säynätjoki, A.; Karvonen, L.; Alasaarela, T.; **Korn**, D.; Alloatti, L.; Tervonen, A.; Palmer, R.; Koos, C.; Leuthold, J.; Freude, W.; Honkanen, S. K.; 'Loss reduction of silicon slot waveguides with ALD-grown thin films', *OPTO SPIE Photonics West - Society of Photo-Optical Instrumentation Engineers (OPTO-SPIE'12)*, San Francisco (CA), USA, Paper 8266-11 Jan. 21-26, 2012.

[C22] Leuthold, J.; Freude, W.; Koos, C.; Alloatti, L.; **Korn**, D.; Palmer, R.; Brosi, J. M.; 'Slotted Photonic Crystal Slow Light Modulators', *Integrated Photonics Research, Silicon and Nanophotonics (IPRSN),* 2011 paper: JTuA3 Slow and Fast Light (SL) Topical Meeting of the OSA, Toronto, Canada, June 2011.

[C23] Koos, C; Alloatti, L.; **Korn**, D.; Palmer, R.; Vallaitis, T.; Bonk, R.; Hillerkuss, D.; Li, J.; Bogaerts, W.; Dumon, P.; Baets, R.; Scimeca, M.L.; Biaggio, I.; Barklund, A.; Dinu, R.; Wieland, J.; Fournier, M.; Fedeli, J.; Freude, W.; Leuthold, J.; 'Silicon nanophotonics and silicon-organic hybrid (SOH) integration', *General Assembly and Scientific Symposium, 2011 XXXth URSI,* Istanbul August 2011.

[C24] Koos, C.; Alloatti, L.; **Korn**, D.; Palmer, R.; Hillerkuss, D.; Li, J.; Barklund, A.; Dinu, R.; Wieland, J.; Fournier, M.; Fedeli, J.; Yu, H.; Bogaerts, W.; Dumon, P.; Baets, R.; Freude, W. and Leuthold J.; 'Silicon-Organic Hybrid (SOH) Electro-Optical Devices', *Optical Society of America IPRSN,* 2011, Paper IWF1.

[C25] R. Palmer, L. Alloatti, D. **Korn**, M. Moosmann, K. Huska, U. Lemmer, D. Gerthsen, T. Schimmel, W. Freude, C. Koos, and J. Leuthold, “Smooth and ultra-precise silicon nanowires fabricated by conventional optical lithography,” in *CLEO:2011 - Laser Applications to Photonic Applications*, 2011, p. CThZ1.

[C26] Freude W., Alloatti L., Vallaitis T., **Korn** D., Hillerkuss D., Bonk R., Palmer R., Li J., Schellinger T., Fournier M., Fedeli J., Bogaerts W., Dumon P., Baets R., Barklund A., Dinu R., Wieland J., Scimeca M. L.; 'High-speed signal processing with silicon-organic hybrid devices', *European Optical Society Annual Meeting (EOS'10)*, Parc Floral De Paris, France, October 2010.

[C27] Alloatti L., **Korn** D., Hillerkuss D., Vallaitis T., Li J., Bonk R., Palmer R., Schellinger T., Barklund A., Dinu R., Wieland J., Fournier M., Fedeli J., Bogaerts W., Dumon P., Baets R., Koos C., Freude W., Leuthold J.; 'Silicon High-Speed Electro-Optic Modulator', *Group IV Photonics 2010*, China, paper ThC2 Sept. 2010.

[C28] Alloatti L., **Korn** D., Hillerkuss D., Vallaitis T., Li J., Bonk R., Palmer R., Schellinger T., Barklund A., Dinu R., Wieland J., Fournier M., Fedeli J., Dumon P., Baets R., Koos C., Freude W., Leuthold J.; '40 Gbit/s Silicon-Organic Hybrid (SOH) Phase Modulator', *ECOC 2010*, Torino, Italy, Paper Tu.5.C.4, September 2010.

[C29] Leuthold J., Koos C., Freude W., Vallaitis T., Alloatti L., **Korn** D., Dumon P., Bogaerts W., Baets R., Biaggio I., Diederich F.; 'All-optical signal processing with silicon-organic hybrid slot waveguides', *OSA Integrated Photonics Research, Silicon and Nano Photonics (IPR)*, Monterey (CA), USA, Paper IWC1, July 2010.

[C30] Freude W., Leuthold J., Alloatti L., Vallaitis T., **Korn** D., Palmer R., Koos C., Brosi J.-M., Dumon P., Baets R., Scimeca M. L., Biaggio I., Breiten B., Diederich F., Barklund A., Dinu R., Wieland J.; '100 Gbit/s electro-optic modulator and 56 Gbit/s wavelength converter for DQPSK data in silicon-organic hybrid (SOH) technology', *Proc. IEEE Photonics Society (formerly: Lasers and Electro-Optics Society) Summer Topicals 2010 "Novel Waveguiding, Structures and Phenomena"*, Playa del Carmen, Riviera Maya, Mexico, Paper WB2.1, July 2010.

Patents

[P1] EP patent application, L. Alloatti, J. Leuthold, W. Freude, C. Koos, D. Korn, and R. Palmer; 'Electro-optical device and method for processing an optical signal', *EP 20110003562, EP 2518556 A9*, Publication date: 2012-12-19.

[P2] USA patent application, L. Alloatti, J. Leuthold, W. Freude, C. Koos, D. Korn, and R. Palmer; 'Electro-optical device and method for processing an optical signal, USA', *US 13/460,395, US 20120321240 A1*, Publication date: 2012-12-20.

[P3] German patent application, L. Alloatti, J. Leuthold, W. Freude, C. Koos, D. Korn, and C. Weimann; 'Wellenleiter-Bauteil für nichtlinear-optische Prozesse zweiter Ordnung', *DE 102012016328 A1*, Publication date: 2014-02-20.

Curriculum Vitae

Dietmar Korn

Dietmar.Korn@gmail.com

Work Experience

05/2009 – 01/2014 **Karlsruhe Institute of Technology (KIT), Institute of Photonics and Quantum Electronics (IPQ), Germany**

PhD studies

- Conception and realization of integrated silicon-organic hybrid electro-optical modulators for advanced modulation formats
- Conception and realization of integrated silicon-organic hybrid lasers

Engineering work

- Adaptation of research work to requirements of industry projects
- Consortium coordination of EU FP7 project www.sofi-ict.eu

03/2007 – 04/2007 **Ferdinand-Braun-Institute Berlin, Germany**

Internship

Characterization of semiconductor laser diodes

04/2004 – 08/2005 **University of Potsdam and Ferdinand-Braun-Institute Berlin, Germany**

Student research assistant

Project 'Strain and tilt analysis of semiconductor disc lasers by means of synchrotron radiation'

Education

05/2009 – 10/2013	**Karlsruhe Institute of Technology, Germany** **Graduate school Karlsruhe School of Optics & Photonics** PhD thesis in electrical engineering, including program for project, human resources management, basic accounting
01/2011	**Vrije University Brussels, Belgium** Intensive training on entrepreneurship in photonics
10/2003 – 04/2009	**University of Potsdam, Germany** Diploma in physics Nonlinear & quantum optics, nonlinear dynamics and computer science
09/2005 – 05/2006	**Université Henri Poincaré in Nancy, France** Erasmus program Graduate studies in physics